D0205052

Valueware

Recent Titles in
Praeger Studies on the 21st Century

The Foresight Principle: Cultural Recovery in the 21st Century
Richard A. Slaughter

The End of the Future: The Waning of the High-Tech World
Jean Gimpel

Small is Powerful: The Future as if People Really Mattered
John Papworth

Disintegrating Europe: The Twilight of the European Construction
Noriko Hama

The Future is Ours: Foreseeing, Managing and Creating the Future
Graham H. May

Changing Visions: Human Cognitive Maps: Past, Present, and Future
Ervin Laszlo, Robert Artigiani, Allan Combs, and Vilmos Csányi

Sustainable Global Communities in the Information Age: Visions
from Futures Studies
Kaoru Yamaguchi, editor

Chaotics: An Agenda for Business and Society in the Twenty-First Century
Georges Anderla, Anthony Dunning, and Simon Forge

Beyond the Dependency Culture: People, Power and Responsibility in
the 21st Century
James Robertson

The Evolutionary Outrider: The Impact of the Human Agent on Evolution
David Loye, editor

Culture: Beacon of the Future
D. Paul Schafer

Caring for Future Generations: Jewish, Christian, and Islamic Perspectives
Emmanuel Agius and Lionel Chircop, editors

Rescuing All Our Futures: The Future of Futures Studies
Ziauddin Sardar, editor

Valueware

Technology, Humanity and Organization

CHRISTOPHER BARNATT

Praeger Studies on the 21st Century

Westport, Connecticut

Published in the United States and Canada by Praeger Publishers
88 Post Road West, Westport, CT 06881.
An imprint of Greenwood Publishing Group, Inc.

Printed in the United States of America

The paper used in this book complies with the
Permanent Paper Standard issued by the National
Information Standards Organization (Z39.48–1984).

10 9 8 7 6 5 4 3 2 1

English language edition, except the United States and Canada,
published by Adamantine Press Limited, Richmond Bridge House,
417–419 Richmond Road, Twickenham TW1 2EX, England.

First published in 1999

Library of Congress Cataloging-in-Publication Data

Barnatt, Christopher, 1967–
      Valueware : technology, humanity, and organization / by
   Christopher Barnatt.
          p.   cm.—(Praeger studies on the 21st century, ISSN
   1070–1850)
      Includes bibliographical references (p.).
      ISBN 0–275–96714–X (alk. paper).—ISBN 0–275–96715–8
(pbk. : alk. paper)
      1. Information technology—Social aspects.   2. Information
   technology—Forecasting.   I. Title.   II. Series.
   T58.5.B37   1999
   303.48′33—dc21   99–24163

Library of Congress Catalog Card Number: 99–24163

ISBN: 0–275–96714–X Cloth
      0–275–96715–8 Paperback

Copyright © 1999 by Adamantine Press Limited

In the hope that we may continue to learn and love, question and dream, even under a digital sky . . .

Also in the Future Trilogy by Christopher Barnatt

Cyber Business: Mindsets for a Wired Age

Challenging Reality: In Search of the Future Organization

For further information see
www.CREaction.demon.co.uk

Contents

Preface

WHAT EXACTLY IS VALUE? And how, in future, will value be created?

This book sets out to answer the above broad questions from a technological, social and organizational perspective. In doing so, *Valueware* looks ahead to the bright and dark futures on offer at the dawn of the 21st century. It also seeks to explore how, in a consumer-driven world that constantly demands further 'progress' and 'profit', we may avoid becoming *technology rich but value blind*.

In many ways, the value journey presented in this book is my attempt to reflect how humanity *feels* about itself today, and how such a feeling may guide all of us in seeking a future sense of purpose and self-worth. Indeed, the aim of what follows is to make you think about business, technology, society, and yourself, in a way that may impact positively on your actions and reactions in the future.

Like most authors to whom the label 'futurist' may be applied, I have become used to the protestations of those many critics who argue that futurology is a waste of time because it is 'never right'. Indeed, when younger I would probably have endorsed their case. However, the more years I see turn, the more I believe that those who complain that studies of the future are 'never right' are somehow missing the point.

Almost certainly, the real purpose of attempting to look ahead is to try and foresee new questions, rather than to concoct any particular 'right' or 'wrong' answers. Like most futurists, I do on occasions present some very solid visions of things to come. However, I do so only to provide icons and mental tools with which to think and plan ahead. We live on a radically evolving planet, yet in the main within a society and business world with its head buried in the sand in fear of change.

The search for value and the 'valueware' that may create it is an enigma that has haunted humankind since we began to think and worry about tomorrow. Ultimately, it is and will remain a sun-chasing quest in pursuit of a moving target that will always set before our brief time expires. Certainly it is a crusade that will never yield a definitive result, and which is hence doomed in the eyes of my aforementioned critics.

In spite of the above, the search for value remains that battle for passion, for belonging, and for a continued presence, to which it is worth dedicating at least some of one's time. Indeed, if I have learnt anything whilst writing this book, it is that the search for value is the greatest grail trek for anything and everything that may further open, comfort or satisfy one's mind.

Christopher Barnatt

Acknowledgements

WRITING THE ACKNOWLEDGEMENTS FOR a book is often problematic. On the one hand, an author tries desperately not to leave anyone out. On the other, one also fights a nagging concern that very long lists of thanks will not inspire the reader. Below I'll therefore try—and fail—to be both comprehensive and brief!

Firstly, having just finished a book on the nature of value and its creation, I would like to thank my parents for being the most influential conscious and unconscious shapers of my own perceptions of the world. Next, thanks wholeheartedly go to those busy individuals who spared the time and thought to contribute to **chapter 6**. Thirdly, thanks to Helen Whalley for once again trawling through a manuscript in order to find so many errors! Also, a nod of appreciation to Sue Tempest for always smiling when I knocked on her door, and to Mark Daintree for listening to so many off-the-wall ideas for too many years to mention.

Finally, I would like to thank Jeremy Geelan for pursuing this book into print. Since persuading me to leap in and say 'yes, contract please!' he has endured an horrific skiing accident, whilst I have merely suffered finger- and brain-pain at the keyboard. Fortunately, we've both pulled through now—as witnessed by the fact that you are holding a completed book in your hands.

Well, I guess the above wasn't either comprehensive or brief. But then little to be said of value ever is.

Prologue
Dawn of the Toolmaker

SLOWLY THE EARTH CONDENSED, its core crystallizing out of cosmic dust and gas. Then gravity really took hold. Rock began to crack, heat, melt and boil. Volcanic eruptions spat fire and vapours through the crust. Some lingered as an atmosphere, whilst others condensed into barren oceans.

Perhaps by chance, a slurry of biochemicals reacted and bacteria emerged. Life had been granted to an already ancient sphere of space aggregate. Much later, single-cell organisms evolved. Far later still, these second-generation lifeforms began to specialize and collaborate.

Eventually, some of the Earth's first composite creatures managed to claw their way out of the waters, and the real conflicts began. For a while tooth pierced hide. Then stone and wood, fire and metals, were adopted as the survival weapons of naked, bipedal flesh and bone. At once the rest of nature didn't stand a chance.

In barely the blink of an eye, the plague spread from pole to pole. Planetary transformation followed. Battles were fought on a greater and greater scale as new weapons continued to replace the old. The fiercest nemesis was science, a pure product of so many hungry and confused brains. Indeed, its instruments of change and annihilation rivalled even the menace of corrupted souls.

And for a while they festered on the brink.

Then somehow, very quietly, the sirens began to sound. Also, largely unnoticed at first, interconnections grew between technologies and across minds.

The Earth paused to lick its wounds and to sigh with age. Evolution was kick-starting once again. They who had been so many were finally becoming one.

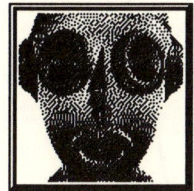

1
Prelude

AS A LECTURER IN computing and organizations, one of the most common questions I am asked is 'what computer should I buy?' It is a query raised by eager undergraduate and postgraduate students—and sometimes even by their parents—as well as by colleagues, family members, and friends. Often, however, my initial response to this 'simple' request is met with more than a little disdain. For rather than reeling off a list of the latest technical specifications, I usually reply by asking 'so what do you want to use this computer for?'

After property and cars, for many households a computer is now their single most expensive possession. Since 1991, businesses have also invested more in computing and communications systems than in traditional manufacturing plant.[1] It is therefore a pity that many people and organizations allow themselves to be guided by sales pitches, rather than by their actual requirements, when making computer purchases. Or to put it another way, when it comes to computing, many individuals and companies seem to have become *technology rich but value blind*.

The majority of computers are fitted with upgrades and peripherals that their owners probably don't need and can often ill afford. Even if this is not the case, then almost certainly the majority of hard disks[2] are now stuffed full with software suites that boast far too many functions for their average user to comprehend. Indeed, it has been estimated that around ninety per cent of the available options in many major programs go untouched by all but the most persistent technofreaks or computer professionals.

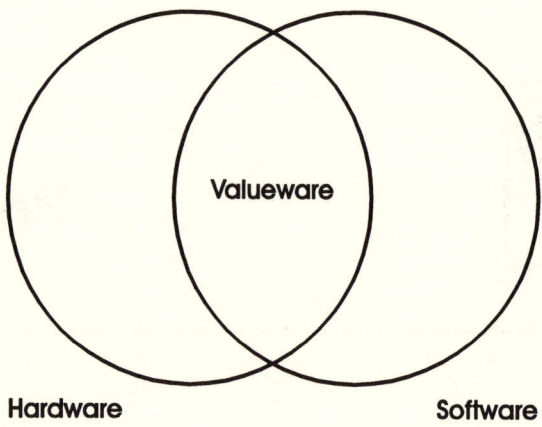

Figure 1.1 Hardware, Software and Valueware

VALUEWARE

The key problem faced by purchasers of computers is that vendors want to sell them hardware and software. Hardware comprises all of the physical components of a computer—such as its display, microprocessor, memory and keyboard. In contrast, computer software is the intangible program code that makes the hardware do something useful. Hence, just as a walkman and a music cassette are useless until brought together, so computer hardware and software are ineffectual except (perhaps) in combination.

The above statement may appear old hat and extremely obvious. However, what it should highlight is how computer purchasers don't really want to buy hardware and software at all. Rather, what both individuals and organizations actually desire is *valueware*—or in other words that product *combination* that will actually do something useful.

Herein, 'valueware' will be defined as only those elements of hardware and software that work in tandem to improve people's lives or business operations. **Figure 1.1** therefore illustrates valueware as existing at the productive interface of hardware and software. What rational computer

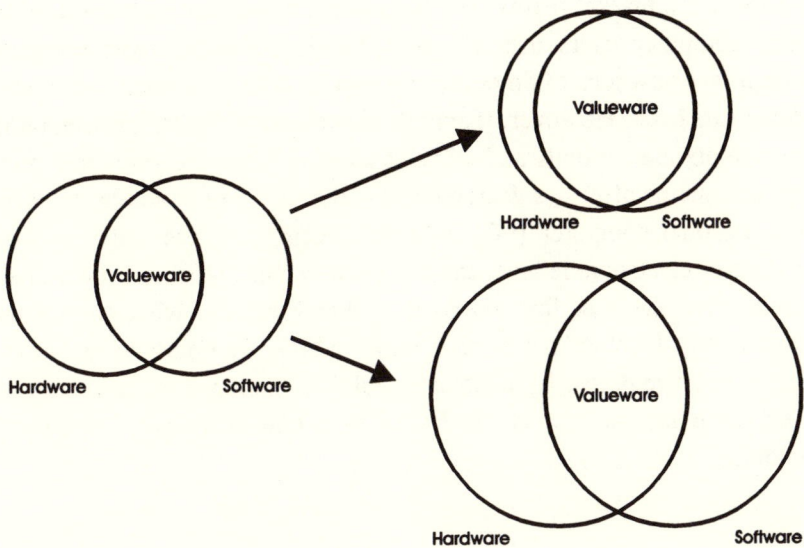

Figure 1.2 Alternatives for Increasing Valueware

purchasers therefore ought to aspire towards is the maximization of this common region.

As illustrated in **figure 1.2**, valueware in computing may be increased in two distinct ways. At one extreme, valueware may be expanded by purchasing the same volume of hardware and software as previously, though ensuring that time and money are only invested in products with a greater value overlap.

Such a situation is shown in the top right of the figure. However, its attainment is often problematic. This is because most computer companies want to sell higher and higher specification hardware and software in order to maximize their revenues. As a result, and as shown in the bottom right of the figure, we are witness to valueware increases achieved not by improving the proportion of the hardware/software overlap. Instead, users are forced to purchase not just more valueware, but in addition the same proportion of 'waste' hardware and software.

Due to this market reality, cost/benefits in valueware have declined for many computer users over the past decade. Granted, most computer systems are now tens of times more powerful than their ancestors of even a few years back. However, if appraised in terms of useful function rather than the number of bells-and-whistles attached, then the *value* that many typical systems offer has decreased in real terms. For example, the price of a personal computer (PC) capable of being used as a decent word processor, electronic ledger, and basic communications tool, could have halved over the past few years. It is therefore somewhat galling that instead entry-level PC prices have remained fairly constant in real terms. This is because perfectly satisfactory 'old' technology has been clutched from the consumer's grasp in favour of a now obligatory multimedia jamboree.

BROADER AGENDAS

The valueware paradigm may prove helpful to have at the back of the mind when considering future computer purchases. However, the concept may also meaningfully be applied in a far broader organizational and human context well beyond the confines of the modern desktop. Indeed, in seeking to answer the two, broad questions of 'what is value?' and 'how, in future, will value be created?', the search for valueware overlaps between technology, humanity and organization is what this book is all about.

Taking an even broader perspective, *Valueware* may also be read as the final volume in my 'Future Trilogy'. Written over a five-year period, this is a series of linked books that report my research and vision for the structure and functioning of both business and wider society in the 21st century.

The first volume in the Trilogy—*Cyber Business*—concentrated almost exclusively on how the emerging technologies of 'cyberspace'[3] and virtual reality (VR) will soon alter our business and personal lives. As a social counterbalance, its sequel—*Challenging Reality*—then investigated how the very *spirit* of humanity and its organizations has evolved since the dawn of civilization. In essence, as the last part of the cycle,

Valueware now ties together the two wide, future agendas of its sister volumes. It also shares their overriding assumption that, on the edge of the third millennium, the human race is facing a step change in its collective evolution.

RADICAL TIMES AHEAD?

My belief that tomorrow will be significantly different from today is, hardly surprisingly, not without its critics. Most often, in opposing the suggestion of any radically different tomorrow, traditionalists argue that the late 20th century is witness to no lesser and no greater a flux of change than any other period in history. In fact, in so dismissing any clear-cut 'post-industrial revolution', some have even suggested that there was never really a step change from land to city during the commonly-accepted timespan of the first Industrial Revolution.[4] Indeed, the case has been made that all popular historical watersheds have been dreamt up after the event by wide-eyed academics in search of neatly-segmented subject areas.

The view that there is no validity in isolating any key periods of historical transition is probably more than a little cynical. This said, it is certainly the case that patterns of civilization have rarely stood still between periods of absolute change. Any pending overnight switch to some dream or nightmare 'reengineered society' therefore has to be dismissed as a foolish proposition.

The above points noted, I still firmly believe that there is something radical and unique about the times in which we now live. Furthermore, within this book I plan to demonstrate how the aftermath of our current challenges and transitions will determine patterns of collective human activity (as well as their valueware) for a very long time to come.

QUESTIONING THE RELIGION OF PROFIT

Before we can progress further in our search for value and the future means of its creation, we clearly need some common understanding of the term 'value' itself. Many economists and management gurus still apparently believe that individual and organizational performance may sensibly be measured in terms of money. As a result, the business world remains focused upon short-term profit creation and maximized shareholder returns. Yet, outside of the plush concrete and glass monoliths that still serve as icons of corporate prosperity, social turmoil and human inequity are increasing. New technologies are also waiting in the wings. These will redefine the nature of business, and may even reengineer humanity itself.

The divide between the 'haves' and the 'have-nots' is becoming ever more stark within nations both prosperous and poor. For an 'underclass' discounted by wider society, subsistence itself is now an increasingly stressful challenge, and happiness often a faraway dream. However, even for those at the other end of the spectrum, life is not as sweet as it once was.

Long-term job security no longer exists. Hours spent at work are also increasing. As a result, many people's personal and family relationships are being strained by new working arrangements from which they feel powerless to escape.

Around the world, 'value' in terms of economic output has doubled in the latter decades of the 20th century. However, the quality of life of many human beings has deteriorated in the same period. These facts alone should serve to highlight how economic measures represent only part of the picture in any study of value and its creation.

As long as we cling to the pleasures and emotions of the flesh, 'value' will never be able to be easily equated to 'profit' or any other form of monetary reward. Certainly, money is something greatly desired by most individuals and organizations. However, this is only because we have collectively decided that balances on paper and in computers represent the best means of allocating the scarce resources of ourselves and our planet.

Money is simply a means to an end that rarely makes people happy in itself. Rather, what human beings and their organizations really value are survival and physical comfort, passion and power, security and stability, love and creativity, wonder and winning, self-fulfilment, and even the dubious elixir of immortality.

Almost inevitably, what is perceived as 'of value' is a very personal thing. However, I would suggest that *value is always something that any individual or organization would prefer to have rather than not have*. Such a straightforward definition of value will therefore be used throughout this book.

TECHNOLOGY, HUMANITY & ORGANIZATION

Since I began to write as a futurist, I have come to believe that three distinct yet increasingly interrelated areas of study should equally demand my attention. The first is *technology*; or in other words the collective nature of all of the tools, programs and media that human beings use at work and at play. The second area of study is then *humanity* itself; or more specifically the nature of what we now perceive as 'human' at this particular point in our social and cultural evolution. Finally there are *organizations*; or those mechanisms into which humanity and technology synergistically invest to form something that may take on a collective life and identity of its own.

The study of the *interrelation* of the above key building-blocks is perhaps the most fascinating if confusing discipline to which anybody may dedicate their time. It is therefore rather a shame that most teachers and academics—and hence most educational curricula—narrowly blinker their studies into just one of these three broad strands. More tragic still is the fact that the vast majority of young people must therefore leave educational establishments having been pigeonholed as 'scientists' or 'social scientists' or 'artists' so early in their lives.

In order to study how technology, humanity and organization may in future evolve and create value, it will prove useful to have a common basis for defining the valueware of each in isolation. Once common valueware definitions have been highlighted, we will also be in a better position to understand how technology, humanity and organization may best combine as *integrative* value machines. The following sections will therefore address the definition of, and similarities between, technological, organizational and human valueware.

TECHNOLOGICAL VALUEWARE

At the start of this chapter, I demonstrated how valueware in computing may be isolated at the hardware/software interface. Opening out our analysis further, it is also possible to define 'technological valueware' more generally as existing in this common area.

Already the computer systems built into an automobile cost more than the metal from which the vehicle is constructed. Programmable production lines are now also involved in manufacturing a great deal of factory output. Services from banking and mail order purchasing, to accountancy, design, and healthcare management, are additionally most often routed through computerized information technology systems. Even domestic appliances, from washing machines to heating systems, toys to automatic garage doors, are today commonly controlled by microprocessor.

In a very real sense, as computer power permeates our lives and environment on so many levels, all of our 'technologies' are rapidly becoming combinations of hardware (physical tools and infrastructures) and software (intangible media content and program code). We may therefore reasonably consider a great many of those technological systems that add value as existing at some productive hardware/software overlap, as previously shown in **figure 1.1**.

ORGANIZATIONAL VALUEWARE

It is, however, not just technological systems that are in the main hardware and software combinations. Organizations, for example, may sensibly be viewed as economic and social machines boasting clear hardware and software components.

In this context, the hardware of an organization consists of its structural blueprint of job roles, contracts, departmental procedures, and regulations. In contrast, organizational software equates to those productive *processes* that may be observed to 'run' across its hardware specification. Effectively, therefore, the agreed *means* for permitting people and technology to carry out tasks constitute an organization's hardware, whilst the activities an organization may actually be observed *doing* constitute its software component. Organizational valueware may

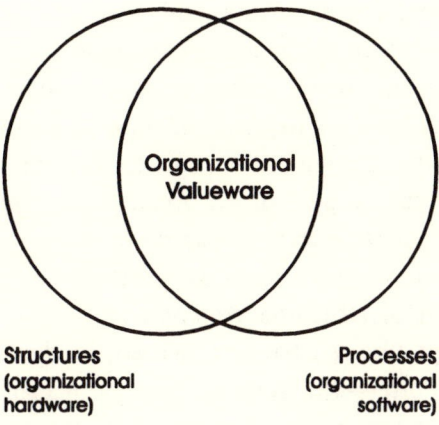

Figure 1.3 Structures, Processes and Organizational Valueware

therefore be illustrated as existing at a productive structure/process interface, as shown in **figure 1.3**.

Viewing organizations as combinations of hardware and software is significant in that it provides a new perspective on the decades-old 'process–structure' debate. This questions whether managers ought to design rigid structures in order to maximize performance, or if instead they should optimize chains of activity *across* their business. In the latter case, loose, flexible patterns of operation will be allowed to emerge and evolve.

For many years, management theorists and practitioners believed that the route to organizational success lay in the adoption of a 'strategy–structure' rather than a 'purpose–process' doctrine.[5] Top managers therefore acted as structural architects whose primary job was to formulate strict systems of control. Job specifications hence became very rigid, with clear rules laid down to govern how every task was to be completed and by which division. From the end of the 19th century, and indeed well into the second half of the 20th, mighty, bureaucratic hierarchies were therefore constructed within which everybody and everything had its proper place and function.[6]

However, in more recent decades, the wisdom of tightly specifying organizational structures has increasingly come to be questioned. As markets have become more turbulent and product lifecycles have shortened, so many have argued that it is now the *processes* of organizations that ought to be management's primary concern.

It is, after all, the outputs of an organization's processes that are desired by its customers. Recent management initiatives[7] have therefore focused upon viewing organizations as 'portfolios of processes', rather than as hierarchies of structural demarcations. This inevitably forces companies to adopt a more customer-focused, market driven mentality. Indeed, Michael Porter—the guru of strategic management—has long argued that organizations need to focus upon the processes inherent in the 'value chains' that actually produce what their customers demand.[8] As a consequence, the functional integration of traditionally-separate departments—such as research and development, production, marketing, and distribution—becomes a key weapon in any process-orientated approach to value creation.

Discounting either of the above extreme views, the valueware concept suggests that neither a structure (hardware) nor process (software) organizational focus is the most sensible. Rather, businesses seeking to optimize their operations ought really to focus upon maximising their valueware overlap. They may do this by ridding themselves of inappropriate structures *and* ineffectual processes.

Unfortunately, many organizations still create rules and bureaucracy for their own sake. A significant number also direct energies into internal activities that add no value to the business whatsoever. Hardware and software resources are therefore being wasted outside of the valueware region.

Much has been written in recent years about so-termed 'virtual organizations' or 'virtual companies'.[9] These are commonly described as having minimized both structural rigidities and in-house resources. Often this has been achieved by applying advanced information technologies to link workers into geographically-diverse 'virtual teams'.

Whilst many interesting cases of virtual organizations have been reported, any solid definition of the phenomenon has yet to emerge. However, utilizing the concept of organizational valueware, I would

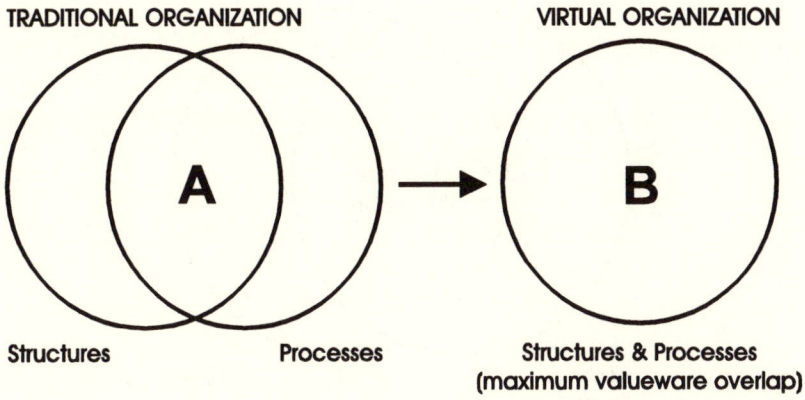

Figure 1.4 **Using Valueware to Define a Virtual Organization**

suggest that virtual organizations are those that have somehow maximized their structure/process overlap. They therefore direct no energies into structures or processes that do not add value.

Virtual organizations may subsequently be distinguished from their more traditional (and less effective) cousins as in **figure 1.4**. Broader discussions concerning how managers may actually move their organizations away from point 'A' and towards point 'B' in this figure take place across the rest of this book.

HUMAN VALUEWARE

In common with technological and organizations systems, even human beings may be pictured as hardware and software combinations. In this context, the body equates to human hardware. Human software then comprises our mind, soul or spirit. That which largely identifies any particular individual may therefore be illustrated as a hardware/software combination as shown in **figure 1.5**.

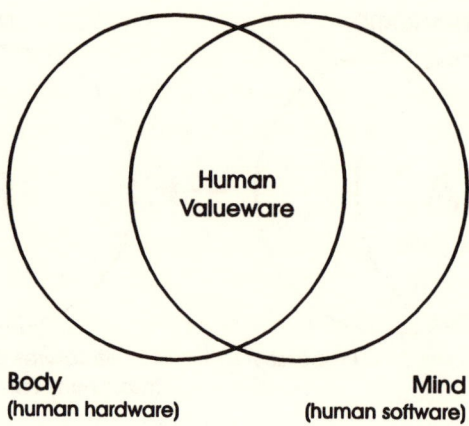

Figure 1.5 Body, Mind and Human Valueware

When considering human valueware, it is not perhaps how the productive overlap of mind and body may be 'improved' that constitutes the most significant area of study. Rather, it is surely more poignant to wonder which 'hardware' and 'software' components need to continue to be present and/or combined for humanity itself to remain intact.

Today, the sciences of computing, engineering and medicine continue to cross astonishing frontiers. As a result, the *technical* opportunities for the replacement of body parts with artificial, cybernetic prostheses continue to grow. Possibilities for linking the human brain—and potentially even the data and programs of the mind—into synthetic inorganic or biotechnological systems will also sooner or later present themselves.

In addition, geneticists continue to promise new skills in the manipulation of our DNA coding.[10] Techniques that may lead towards the potential run-of-the-mill cloning of human beings or parts thereof are also being honed. Hence, as time marches us further into the third millennium, the ability of the human race to alter 'artificially' its own hardware and software will become a commonplace weapon. The very *nature* of future human valueware is therefore likely to be a hotly-debated topic indeed.

Key to much analysis of the fusion of 'human' hardware and software with that of lifeless 'technology' will be the moral and social agenda behind such synthesis. For those facing death or disability, the manipulation of the hardware of their bodies with prostheses that may extend their span and/or quality of life is likely to raise little dispute. However, already some perfectly healthy individuals have suggested that they will seek alterations to their bodies (and in some cases even brains) as soon as certain technologies become available.

For example, some 'transhumanists' have voiced desires for cybernetic limbs that may give them greater strength, microelectronic eyes for a broader range and clarity of vision, and even 'brain interface' skull sockets to allow their memories and mental capacities to be augmented with computer technologies.

Driving such notions of 'transhumanity' are its advocates' beliefs that a proactive stance ought to be taken in 'upgrading' the hardware and software of our own evolution. Of course, to many, such an idea is scary and totally *inhuman*. To lose tonsils, an appendix, or even a limb, certainly does not cause the valueware of anyone's body/mind hardware/software overlap to be questioned. Similarly, being fitted with a pacemaker does not make a person any more or any less human. But what about being interfaced directly with a computer? Or boasting over fifty per cent microelectronic body parts? Or even uploading one's consciousness into a computer network in order to work with financial data at its most basic level?

The above scenarios may sound ludicrous. Yet, as we shall see in **chapter 4**, over the next few decades they will all become very real possibilities as the first 'posthumans' discard the present limits of the flesh. Indeed, even before that day occurs, both individuals and organizations may be alarmed to realize how boundaries between many previously distinct hardware, software and valueware combinations are starting to become more than a little blurred.

A SOFTENING OF FOCUS . . .

Whilst previously defined as discrete, technology, humanity and organization are increasingly partners of mutual interdependence. For a start, all advanced technologies may only be developed and manufactured as a result of the collective actions of human beings within organizations. Organizations in turn can only exist when human beings choose to act cooperatively together and to utilize certain technologies. And today, even humanity itself could not survive without a wide range of technological and organizational mechanisms to sustain the complexity of modern civilization.

Biophysicist Gregory Stock has suggested that technology, humanity and organization are now so closely intertwined that in amalgamation they constitute a new lifeform or 'superorganism'. He has christened this gigantic, new creature 'Metaman'. Stock even suggests that Metaman's recent birth signals the next phase in the evolution of life on our planet.[11]

To adopt such a fascinating viewpoint in an everyday context is undoubtedly somewhat excessive. However, it nevertheless remains the case that technology, humanity and organization are increasingly interconnecting into a thriving whole.

BLURRING INTERFACES

For example, ever since Henry Ford and his contemporaries built the first, great factories of mass production, organizations have been evolving into more and more technological machines. So much is this the case that today millions of people interact with many companies directly through interfaces of plastic, metal, glowing phosphor, and digital audio.

For a decade or more, the bank clerk has been superseded by the cashpoint or 'automatic teller machine' (ATM). When we phone certain organizations we may also end up speaking to a computer. Over the Internet, people now transact with on-line organizations solely by clicking with a mouse. Interactive home shopping TV channels similarly permit purchases to be made by stabbing little keys on an infrared remote control.

Perceptions and practices of 'technology' and 'organization' will continue to mingle as the world electronically joins hands and retreats from inconvenient exchanges in the flesh. Already some individuals have opted to connect to their bank from within a standard *Windows* software package such as *Microsoft Money*. If this technological trend continues, organizations in sectors like financial services may soon face a fundamental shift in their primary public interface from highstreet shop to desktop icon. In turn, companies with potentially substantial on-line markets may be wise to realign their cultures to best serve a *user*, rather than a *customer*, base.

HYBRID LIVES

Just as organizations are mingling with their technologies, so human beings are also starting to mingle with their organizations. For many individuals, new workplace flexibilities are removing previously concrete distinctions between their 'home' and 'business' lives. In parallel, digital technologies are also playing a significant role in diminishing the workplace/homeplace divide.

Computer monitors and television screens, microchips, magnetic storage devices, and a whole host of complex communications systems, are today increasingly divided between our work and our living space. Millions of people labour in offices in front of IBM-compatible personal computers that are identical to those found in at least fifty per cent of Western homes. Telephones, fax machines, and a whole host of other computer-controlled devices exist in a similar social and organizational ubiquity. They are also generally all connected into the same electrical and communications networks.

Fortunately, the above fact does not appear to bother most people. This is probably because we mainly use the common technologies of our working and domestic lives for different purposes. However, what the above point does demonstrate is how many technologies are no longer organizationally or socially specific. Today this may be so true that it is obvious. However, to employees in past times—who did not leave their workplace to return to homes populated with similar technology—it would undoubtedly have proved something of a surprise.

COMPETING WITH SMART MACHINES

The comparative advantages that distinguish human beings from technological organizational resources are also becoming harder and harder to fathom. Most notably, no longer may cold technology always be classed as dumb and task-specific. As a result, human workers are increasingly facing competition as and when businesses reengineer to utilize the most flexible and 'intelligent' factors of production.

With an ever-increasing software component, a growing proportion of productive plant is now multi-functional, programmable and flexible. Perhaps alarmingly, today some computerized systems are also ceasing to be dumb. Indeed, in areas such as credit authorization and geological diagnosis, many machines may now be considered 'smart', if not (perhaps) truly 'intelligent'.

Certainly many machines now have the ability to learn. For tasks requiring adaptability and intelligence, organizations therefore now have a potential choice of employing *either* people *or* machines.

In the other direction, as individual craft skills in most industrial economies have waned, so a great many human beings now find themselves employed as little more than task-specific organizational components. Indeed, the very logic of mass production that drove Western industrialization demanded that the majority should cease to master broad craft skills. It is therefore perhaps hardly surprising that as jobs for many people have been down-skilled, whilst in parallel many machines have 'skilled up', so the productive characteristics of human beings and new technologies have mingled.

Across the globe, interacting computers and other programmable machines are increasingly being trusted with an ever wider variety of tasks. In parallel, many of those people labouring as semi-skilled automata on production lines, in 'office factories', or in sprawling telesales call centres, are finding that all of their latitude at work is being surrendered in the name of consistency and efficiency. Indeed, even those people still blessed with jobs boasting some variety are beginning to find themselves challenged by the inherent flexibility of unthinking and infinitely-reprogrammable silicon brains.

To teach a financial or tax advisor a new set of regulations may take several weeks or even months. However, to update the 'skills base' of an

Internet web site offering a similar service may simply be a matter of changing a few lines of code. Like it or not, it is already sometimes faster and cheaper to re-skill a contraption of metal and silicon than to mentally re-tool an old-style, flesh-and-blood employee.

The threat of a working landscape increasingly dominated by smart technology is now extremely real. However, as we realize this sobering fact we ought not to forget that the division between human beings and technology also continues to blur.

Cybernetic organizational systems reliant upon human beings who conduct disembodied business in cyberspace are already a reality. More and more subtle interface links between flesh-and-bone and plastic-and-metal technology are also in development. For some this second realization may be even more frightening than the first. Yet for others, it represents an opportunity for synthesis and synergy, and incredible new avenues for future value creation.

OUR ROUTEMAP AHEAD

With a blurring of organizational and technological interfaces, the same technologies in use in our homes and workplaces, and human workers increasingly interchangeable with machines, many of the boundaries that once governed our perceptions are becoming indistinct. The study of technology, humanity and organization (and of the value each may create) therefore *has* to become the study of how their *boundaries* are now converging. Indeed, it is my belief that, by studying this coming-together, a greater understanding of how future value will be created is likely to emerge.

Figure 1.6 provides a graphical representation of technological, social and organizational convergence. On the left of the figure, we see three circles representing technology, humanity and organization. Little overlap is shown between them. Such a situation therefore represents a state of the world that increasingly no longer exists. Instead, as shown on the right of the figure, in the non-too-distant future technology, humanity and organization (and their valueware) are likely to share increasingly common domains.

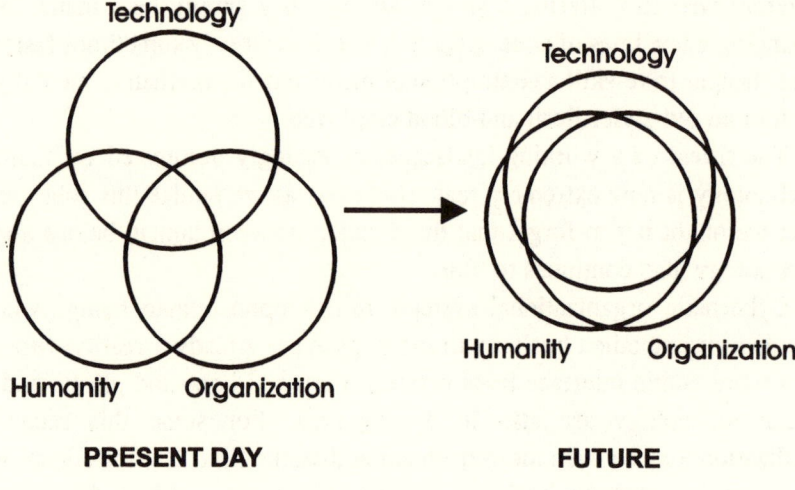

Figure 1.6 Technology, Humanity and Organization Converge

A wide range of industrial and economic analysts have now highlighted 'convergence' as one of the most dominant paradigms of the late 20th century.[12] Most support this proposition by noting how the three key information industries of computing, telecommunications and media/entertainment are now blurring into one. Disputes remain as to whether it will be software giants like Microsoft, network infrastructure suppliers like AT&T, or 'content' providers such as News International, who will triumph in the battle to control any single, resultant digital 'meta-industry'. However, there does seem to be general agreement that the 'convergence phenomenon' is effectively limited to emergent commonalities in computing, telecommunications and the media.

Given our previous consideration of the blurring boundaries between technology, humanity and organization, the above proposition is more than a little surprising. However, the blinkered view of convergence in current debate is almost certainly due to the fact that most of its proponents are either pure information technologists, or else business academics with a keen interest in the definition of traditional industry boundaries.

Far broader—and, in the long-term, far more significant—technological, social and organizational boundary transitions therefore go largely unreported. As a result, the 'real' debate concerning a wide variety of convergence forces upon our lives and organizations has yet to be triggered.

In my own attempt to get this sphere of debate rolling, **Part I** of this book examines three broad sets of 'convergence forces' that are also likely to become some of the most critical value engines of the future. These comprise 'networks and middleware' (as detailed in **chapter 2**); flexible working patterns, (as discussed in **chapter 3**); and those push and pull factors that continue to draw more and more technology into human lives (as debated in **chapter 4**).

VALUE PERSPECTIVES

Once **Part I**'s survey of convergence forces has been completed, **Part II** continues our quest for value, and for the means of its creation, by reporting a cross-section of specific value perspectives. In particular, these middle chapters balance the views of a range of commentators concerning management evolution, on-line culture, and the next customer generation.

Initially, **chapter 5** outlines both the traditional and more cutting-edge approaches to management that have driven, and may (perhaps) continue to drive, business organizations towards corporate success. **Chapter 6** then gleans what we may learn about value and its future creation from a study of those communities of individuals who already inhabit the cyberlands of the Internet. This global ensemble of computer networks has rapidly become the greatest ever info-nexus of humanity. Already it constitutes the very soul of Gregory Stock's Metaman. The Net is therefore ignored by serious researchers in any discipline at their peril.

Finally, before we reach the two concluding chapters of **Part III**, the last set of value perspectives to be considered are those of the 'neXt Generation'. Specifically, **chapter 7** examines the attitudes, skills and mindsets of those young people today who are destined to become the key workers, customers and value-shapers of century 21.

THE TRIUMPH OF PLASTIC

Not long ago I was in a checkout queue in my local supermarket. The month was October, and a new intake of students had just descended upon my university town. As usual, some of these young people had travelled from other countries to further their education in Nottingham. Certain peculiarities of day-to-day British life were therefore bound to be unknown to them.

Before me in the checkout queue was one such young woman. A strikingly-attractive blonde with a mid-European accent, she became more and more nervous as the slow-moving line of shoppers advanced. In particular, I noticed how she constantly kept looking into her purse to fumble the peculiar pounds and pence within. Finally, just before committing to transfer her purchases from basket to conveyor, she turned to me and asked, 'Do you know, do they take cash here?'

It was one of those strange moments when you suddenly realize how rapidly the world keeps on changing. I did of course assure the young woman that the supermarket would accept her cash as payment. However, the fact that she had even *imagined* that physical note-and-coinage would not be acceptable made me think. I hadn't noticed how every customer in the queue before us had only swiped plastic. Yet to her this was clearly unusual.

Back at the start of this chapter, I stated my belief that we are currently undergoing a step-change in many of our current 'realities', and hence that the future will be radically different from the present. Even if this proposition is not the case, what the story outlined above serves to remind us is how radical change is one of the few certainties of modern life. Whether evolving on a continual or step basis, the future *is* destined to be very different from the world of today.

Aldous Huxley once stated that experience is not what happens to you, but rather what you *make* of what happens to you. What this philosophy clearly implies is that we all have a choice in how we govern and expand our perceptions of the world.

How much you and I may learn about the future has to depend upon the effort we are prepared to put into understanding the changing world of today. Towards this end, much of this book is concerned with an examination of small and 'obvious' changes in our lives that, upon further analysis, are really quite radical.

To the disappointment and anger of many, the future will not explain itself as it arrives. Nor will it be carefully presented to us in distinct and manageable chunks. The meanings and valueware of the future are therefore something that we all need to start thinking about today. Indeed, true 'value' has often been found most clearly by looking inwards to the soul, rather than outwards to the 'traditional' world in which so many of us still *think* we live.

PART I
CONVERGENCE FORCES

2
Networks & Middleware

SOMETHING IS HAPPENING TO computer technology and business organizations, and that something may be described as 'connectionism'. Like it or not, 'getting wired' has become the greatest business and social phenomenon of our age. Most computers are being linked to each other. Many businesses large and small are also increasingly dependent upon a broader and broader network of partner organizations.

In parallel, customers around the planet are getting used to locating and purchasing many products and services via communications and computing technologies already common in the home. Already, to be 'wired' is not just to have an Internet connection, but to have adopted a 'networking mentality' for linking ideas and traditional economic resources in new ways. Something is happening to computer technology and business organizations, and it is a symbiotic something rather radical indeed.

As connectionism advances and 'being wired' becomes the norm even for those *without* a mobile phone, so many previously certain boundaries between discrete technologies and organizational infrastructures look set to blur. In order to examine this transition, this chapter first details how advancements in computer networks and new 'middleware' are interacting to create a new business environment. Analysis then broadens to examine how increased technological and organizational connectivity are now impacting upon the mindset required for effective management.

Sadly, you will frequently hear the Internet dismissed as a fad. In particular, many believe that its on-line frontier is marred by pornography, technobabble, content of dubious integrity, and far-too-few business opportunities. However, even if all of these questionable propositions are accepted, the Internet still remains the first fully-interactive global phenomenon. Browsing its ocean of tens of millions of words and images, anybody with wired computer access may now touch more of humanity than previously possible in even the richest of past lifetimes.

Today the Internet—the largest functioning anarchy in the world—presents us with an extraordinary, clinically-digital gateway to our emerging and converging technoculture. With it, those of us interested in the future may finally begin to scratch our evolving, collective soul. Indeed, it is worth remembering that people with Internet access are not really wired to computers, but to one other.

INTERPERSONAL COMPUTING

Over the past decade the computer hardware that still consumes many a modern briefcase or desktop has evolved from 'personal' to 'interpersonal'.[13] As a result, computers now have a dual role as information processing *and* communications devices. Indeed, for many people, the sole purpose of a computer these days is to act as a communications conduit. Many times, for example, have I heard the tale of a previously technophobic grandparent who invests in a PC and Internet connection as soon as their grandchildren get on-line.

Visions and sometimes even plans for a new, high speed 'information superhighway' or 'global information infrastructure' (GII) continue to spawn. However, it is the humble anarchy of the Internet that looks set to continue to drive the on-line revolution. Even Bill Gates' *Microsoft Network* (MSN) failed to challenge the Internet's power, and quickly became just one of many services available over its rambling, global span. Certainly, higher capacity and more reliable network systems will continue to emerge. However, most look equally certain to be interfaced at some level with the Internet, or at the very least with its more recent 'intranet' or 'extranet' offspring.

The Internet itself is best defined as the world's largest, public-access collaboration of autonomous yet interconnected networks. The system is based upon an open communications standard known as 'TCP/IP' (the transmission control protocol/Internet protocol). This was first developed for the US military's ARPAnet[14] back in 1969. Since then, TCP/IP has been adopted by a wide range of other research, education and business networks.

The Internet creates value from two killer applications. The first is electronic mail (e-mail), which permits rapid and very-low-cost electronic communications. The second is the world-wide web (WWW, or simply 'the web').

The web has become synonymous with the Internet for many users. Accessed via a program known as a 'browser', it provides a graphical interface to hundreds of thousands of pages of on-line resources. This is achieved by 'hotlinking' text and graphics upon 'web sites' so that visitors need only to click with a mouse in order to interact with the page, or to 'surf' to another location.

HOSTS, CLIENTS & ISPs

The hardware of the Internet can be divided into three basic elements. These comprise its communications infrastructure; the powerful, 'host' computers that store content and route data through the network; and finally a much larger number of end-user computers known as 'clients'.

Every autonomous network that forms part of the wider Internet will have at least one Internet host computer. All hosts have a permanent network connection (also known as a 'backbone link' or 'gateway') to at least one other external host. This array of permanent host-to-host connections thereby maintains the *interconnectivity* (rather than direct connectivity) of all the individual networks that in amalgamation comprise the entire Internet.

Whilst all Internet hosts are permanently interconnected, most Internet users do not enjoy direct access to a host machine. They therefore have to connect in via a 'client' computer (typically a PC) when they want to access the Internet and its host services. In order to obtain a client connection, most Internet users purchase an account with an Internet service provider (ISP) or 'PoP' (point of presence), which then enables them to connect to this organization's host computer.

Internet clients generally connect to their host/ISP over the public telephone network. The vast majority of the network infrastructure over which Internet communications travel is therefore not dedicated to the system. This 'piggybacking' of the Internet over the public telephone network in part accounts for the connection problems of which many client users so frequently complain.

Because not all hosts are connected to all others, most Internet communications have to be 'routed' between several hosts to complete their journey. This is in addition to communications having to travel between hosts and clients, and again can slow the system down.

Figure 2.1 provides a diagram representative of the basic hardware of the Internet. It also indicates the possible pathway of a typical message from PC Internet client *'User A'* to its host, from host to host across the permanent backbone infrastructure, and finally on to its intended client recipient *'User B'*.

GETTING BUSINESS ON-LINE

Today most businesses seem keen to court an Internet presence. Often this is in the hope of *future* network-based value creation. However, some retail organizations—including Amazon.com, the Innovations Group, Value-Direct, and the supermarket Tesco—are already selling large product ranges on-line.[15]

A plethora of other firms are also successfully using the Internet to improve their service level. Usually this involves offering on-line catalogues, product information, and/or after sales support. Many such companies are admittedly in the computer industry. However, the use of the Internet's graphical world-wide web as a shop front for organizations from estate agents to jewellery makers, charities to local government bodies, is continuing to increase. Indeed, a few firms have already turned interactive, on-line service level improvements into key product differentiators.

As just one example, United Parcel Services (UPS) now boasts a web site that enables any customer to instantly track the location of their shipment any time and anywhere in the world. In addition to enhancing service quality, this saves UPS a great deal of money. Indeed, the cost of answering a typical parcel information request via the UPS web tracking system is only seven cents. This compares to a price tag of around two dollars when the same customer information is supplied over the phone.[16]

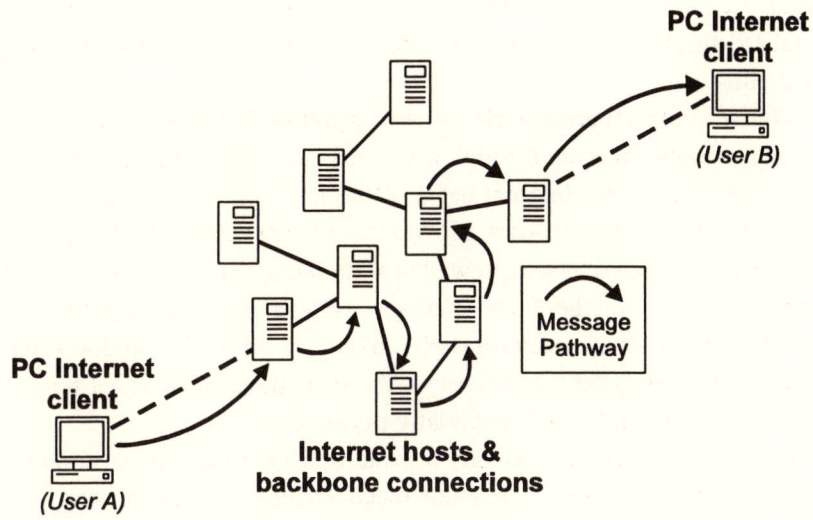

Figure 2.1 A Conceptual Schematic of the Internet

Notwithstanding these examples, many people remain sceptical of the actual or potential business benefits to be reaped from the Internet. In particular, legitimate fears continue to be raised concerning the slow speed and insecurity of the network for some business purposes. Many firms wishing to benefit from Internet technologies, but wanting to avoid such headaches, have therefore turned to the development of private 'intranets' and 'extranets' in their bid to reap most value from network connectivity.

THE INTRANET REVOLUTION

Intranets are internal, private networks that use the TCP/IP communications standard and open software of the Internet within a single organization or part thereof. The easiest way to conceptualise an intranet is therefore as a closed, company-specific world-wide web. This said, intranets may be used for a wide variety of network functions from archive and multimedia document access, to e-mail communications, database distribution, video-conferencing, and even on-line virtual reality modelling.

Whilst being inherently closed, many intranets feature a link to the broader, public Internet through a security filter known as a 'firewall'. These usually permit internal users of the intranet fairly open access to Internet world-wide web sites. However, external access into the intranet network is restricted by the firewall to authorized *and authenticated* users only.[17] Workers in the field—perhaps using a portable computer and mobile phone—may therefore obtain access to their organization's internal intranet system. However, at the same time, the intranet remains protected from public and potentially prying eyes.

Because intranets are usually run across an organization's internal network hardware, they are often much faster than the Internet from whose technologies they spawned. Due to speed and security advantages, the sensitivity and volume of material that may be communicated over an intranet is usually far greater than that considered desirable over the public Internet.

The vast majority of large organizations are now developing intranet networks to ease the flow of internal administration, and to improve their knowledge management.[18] Due to their use of standard Internet software tools and interfaces, intranets prove easy to learn for anybody familiar with electronic mail and the world-wide web. Intranets may also enable both individuals and organizations to begin to overcome 'information overload'. This is achieved by facilitating a mindset shift from 'push' to 'pull' network communications.

Many current electronic communications tools—and most notably electronic mail—are based on a broadcast model whereby information producers make the decision as to who may want or need to receive their data. They therefore 'push' their labours out to others as they see fit. For example, by e-mail, a report or letter may be copied effortlessly to ten or twenty colleagues. This is usually regardless of whether all want or need to receive it.

However, under an intranet model, producers of information more commonly place reports, letters and databases upon their intranet site, from where those with an interest will 'pull' whatever data they require. Intranets hence encourage a demand- rather than supply-led information and knowledge management mentality. As a result they may vastly reduce the time and money wasted on 'junk' e-mail.

EXTRANETS EMERGE

Another benefit of devoting computing resources to intranet development is the resultant shift from closed to open network systems. By definition, intranets utilize the TCP/IP and widely-available software tools (such as web browsers) developed for the Internet. By moving towards an intranet/Internet network architecture, companies are therefore fostering their potential to interconnect with other organizations at both a techno-logical and business level.

The recent emergence of 'extranets' clearly demonstrates the flexibility enjoyed by those companies with intranets who wish to engage in electronic business-to-business trade and communication. Extranets are formed when two or more organizations choose to make secure and usually high-speed links between their respective intranets. The companies concerned therefore benefit from an even more widespread use of their open intranet/Internet technologies. They also again avoid the potential problems of communicating through the oft-unreliable, slow, and potentially insecure public forum of the Internet.

An emerging three-tier intranet-Internet-extranet model for various levels of public and private business communication may therefore be represented as in **Figure 2.2**. Here, the totally open, public arena of the Internet sits in the centre as a potential advertising space, product-support forum, and shop window. However, the two companies shown also have the benefit of their own, private intranets for internal communications. Additionally, a shared, private extranet has been mutually created for the interchange of critical business data.

The broader implications of this new technological/organizational business infrastructure will be considered later in this chapter. However, it is worth stressing at this point that organizations which choose not to develop intranet systems are not only missing out on potential internal administrative savings. More significantly, they also risk getting left behind in business-to-business extranet development by not evolving their internal systems to an open intranet/Internet network architecture.

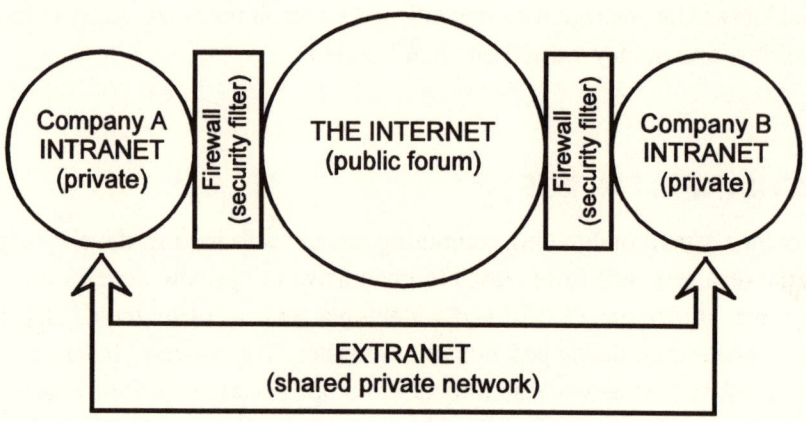

Figure 2.2 The Internet, Intranets and Extranets

MIDDLEWARE HORIZONS

Like the IBM PC before it, the Internet has enjoyed increasing popularity because it has been built upon public, open standards. For example, all of the Internet's key software technologies have been developed in parallel for use on almost all available computer platforms from IBM PCs to Apple Macs and beyond. Because of this, Internet software standards and applications, such as TCP/IP and web browser programs like *Netscape Navigator* or *Internet Explorer*, have sometimes been labelled 'crossware'.[19]

However, in addition to the Internet's own crossware, a broader categorization of computer software known as 'middleware' is becoming equally critical if the most value is to be reaped from any computer or network thereof. As explained by Gil Amelio, the former CEO of Apple Computer, future value-added developments in software technology will not derive from advancements in 'low-level' operating systems or user interfaces such as *Windows*. Nor will they primarily be associated with more advanced 'high-level' end-user applications such as better word processing or database packages. Rather, middleware is fast becoming the new computing and telecommunications software battleground.[20]

By middleware, Amelio is referring to a broadening range of generic software technologies that sit somewhere between operating systems and end-user applications. Already the crossware middleware of the Internet has allowed millions to reap value extremely effectively from open standards in computing and telecommunications connectivity. Indeed, before the introduction of the 'hypertext markup language' (HTML) used to write pages upon the world-wide web, incompatibilities between documents created on different computer systems were rife. Yet, in only a few years, the use of HTML as a transparent middleware has enabled people to forget about vastly-complex network technology, and instead get on with useful multimedia exchange.

Already the crossware middleware of the Internet is blurring technological and organizational boundaries as cyberspace becomes a common social and business medium. However, the burgeoning development of a new range of middleware represents a far more significant force for technological and organizational convergence—and hence increased value creation—upon a great many levels.

Before long, middleware will comprise the glue that will hold businesses to their customers, to their suppliers, and to their partner organizations. This is because new middleware is due to mask much of the inherent complexity of increased technological *and organizational* networking. New middleware software technology will therefore assist enormously in the fluid interlinkage of people and organizations to each other.

THE NETWORKED ENTERPRISE

Just one of the companies intent on middleware creation is the Netscape Communications Corporation. In a recent paper, its Product Development Team outlined its vision for the seamless interlinkage of corporate intranets, extranets, and the public Internet.

Netscape's development of totally open network software standards is intended to permit the emergence of 'Networked Enterprises'. Such companies will use their combined intranet-Internet-extranet technologies to link *directly* with all of their partner organizations, suppliers, distributors, and even domestic customers. These businesses will subsequently begin to take on networked architectures that will mirror those of their interconnected information technology systems.

According to Netscape, the hardware-independent software and communications standards of the Networked Enterprise will provide two primary benefits. Firstly, by creating new types of services, they will allow businesses to engage customers in a directly-interactive fashion that will result in stronger customer relationships. Secondly, by drawing all manner of business partners and customers into product design and evaluation processes, common intranet-to-extranet software will permit companies to bring their products and services to market more quickly.[21]

In the customer's eyes, both of the above developments will blur the boundaries between 'technology' and 'organization'. Indeed, the day is already not far away when *Microsoft Word* and your favourite e-mail package will appear in a software window next to icons for your bank, travel agent, on-line newspaper, and local supermarket. What's more, perhaps less than a decade from now, the integration of desktop software applications with business services may have blurred to such a level that neither remains distinct.

As Don Tapscott argues, 'interenterprise computing' is already empowering electronic business disintermediation.[22] By this he means that customers who choose to interface with organizations over digital networks no longer need wholesalers, retailers, resellers, dealerships or distributors to broker or support their purchases. *Middleware is therefore starting to replace the middleman.* What's more, by removing the intermediary links in many purchasing chains, future middleware may also permit the metabolism of business to shift into real-time.[23]

With learning middleware interfaces, organizations will also be increasingly able to acquire highly cost-effective, specific and actionable customer intelligence.[24] For example, the advancement of world-wide web browsers that save 'cookies' of customer web site interaction details upon every user's PC is just one development poised to deliver a new generation of marketing information.[25] By accessing such 'lumps of code', organizations will not only be able to know how every actual and potential customer has enquired about their services. In addition, they will also have the opportunity to learn how every actual and potential customer has investigated and traded with their competition and in what manner.

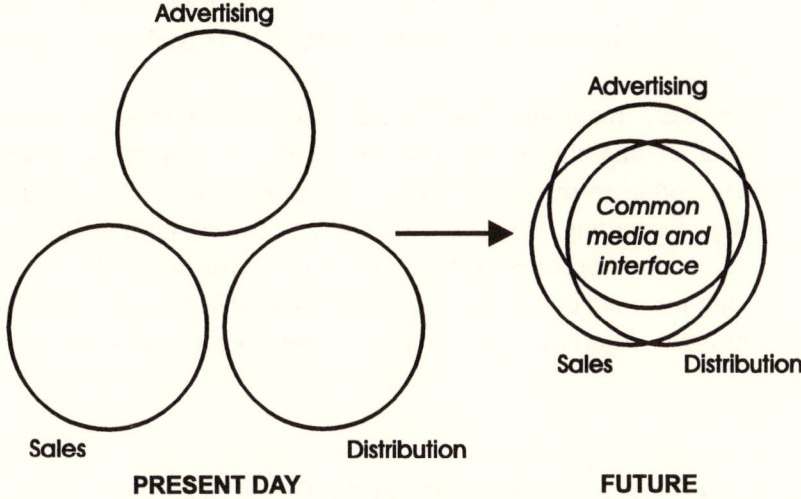

Figure 2.3 The On-line Convergence of Business Functions

BLURRING DEMARCATIONS

A further implication of the overlay of computer network hardware with new middleware software is the likely convergence of many traditional business functions. For example, when roaming the Internet, future customers will click on world-wide web advertisements with which they will then interact as a sales medium. In turn, some future 'digital products' (such as video-on-demand, or legal and financial services) will be dispatched electronically via the same network hardware and point-and-click software interface used for advertising and selling. As a consequence, in the customer's eyes, the boundaries between distinct media and interfaces for 'advertising', 'selling' and product/service 'distribution' will disappear.

Figure 2.3 illustrates the on-line convergence of media and interfaces that many organizations will undoubtedly face over the next few years. For networked enterprises, no longer will adverts be passive sheets of paper or one-way on-screen entertainments. Rather, they will become a two-way middleware, ever poised not just to interact with customers as sales tools, but in addition programmed to deliver back customized, digital product.

The business implications of this inevitable consequence of moving advertising, selling and digital delivery on-line are clearly enormous. However, within many organizations, the inter-departmental political wranglings concerning which division ought to control on-line development may perhaps be even greater. The thought that those who design advertising may, almost by default, end up controlling the on-line sales and distribution interface, may for some prove quite a shock!

AUTONOMOUS MIDDLEWARE SERVANTS

To date, most middleware developments—and in particular those of the Internet—have been concerned with increasing the ease and even friendliness of computer use and/or networked communications access. Such advancements are also likely to continue. In particular, as discussed in **chapter 4**, a shift may well occur away from 2-D desktops and towards 3-D virtual reality (VR) interfaces for the fluid and 'natural' control of many advanced technologies.[26]

Already 3-D VR sites are springing up on the world-wide web.[27] Teleconferencing and telecommuting in VR—known as 'virtucommuting'—are also soon likely to become available to those companies whose networks already facilitate 'remote' meetings via video-conference links.[28]

The above noted, there is more to middleware development than merely interface advancements and the establishment of common standards for interactive, on-line communications and trading relationships. In addition, 'smart' middleware is in development that will permit more complex yet more fluid patterns of interpersonal and customer/organization interaction.

Back in 1968, Arthur C. Clarke wrote *2001: A Space Odyssey*. In this science fiction classic he told how, in the January of 1997, an artificial intelligence known as 'HAL' would be 'born' to pilot the transplanetary spaceship *Discovery*.

By now we know that Clarke's prediction did not come to pass. However, in the January of 1997, the National Aeronautics and Space Administration (NASA) did issue a press release announcing the development of a highly advanced artificial intelligence system for spacecraft control. Named 'Remote Agent', this software will fly the unmanned space probe *Deep Space One* on an exploratory mission. It will also operate with minimal human assistance. Indeed, NASA claims that Remote Agent will be able to reason the state of its spacecraft, as well as having the ability to consider the consequences of its self-selected actions.[29]

The announcement of Remote Agent is significant in that it represents a watershed in the application of smart 'software agent' technology. Software agents are pieces of self-contained computer code that may be instructed to carry out electronic tasks for their human masters. Such 'electronic servants' have been in development by computer scientists for over twenty-five years. However, it is only now—with widespread computer networking based on open standards—that opportunities exist for software agents to create real business and consumer *value*. Indeed, many predict that software agents will soon be 'working' as electronic shopping assistants, researchers, credit-authorizers, fault-finders, travel agents, time schedulers, and more. UK technology consultants Ovum have predicted a market for software agents to be worth $2.6bn around the year 2000.[30]

As the emerging 'working class' of cyberspace, most software agents will possess a mix of up to four basic qualities. Firstly, most will be *mobile entities* capable of travelling around computer networks—and hence into and out of different organizations—in order to carry out their assigned tasks. Secondly, like NASA's Remote Agent spacecraft pilot, most agents will be programmed with a degree of *autonomy* in their decision making patterns. Linked to the latter attribute will be the ability for most agents to *learn*—both from their environment and previous 'experiences', as well as from the actions of their human masters. Finally, in order to prove effective in transparent value creation, most agents will have the ability to *communicate and to exchange data with other agents* (and hence with the companies and individuals whom other agents will represent).

Agents programmed with some of the above qualities are already in application. For example, Apple Computer's 'V-Twin' may be dispatched around a range of networks to find information that's *like* a whole host of other text. It will then produce a summary report of its findings condensed to any length its master may require.[31]

Meanwhile, the 'BargainFinder Agent' created for Andersen Consulting at the Centre for Strategic Technology Research can locate the best on-line purchases for pop and rock CDs. Similarly, the Internet Bookshop's personal librarian 'Jenny' may be asked to keep you informed of new books in which you may be interested. Students at the MIT Media Lab may even enrol on courses where the 'NewT' agent will scan the Internet for an updated reading list every few hours.[32]

Looking further into the future, increasingly sophisticated software agents are likely to be granted more and more autonomy in a business world in which people would otherwise drown in complexity. Most significantly, software agents are destined to remove many of the transactions costs that currently exist between organizations and customers thereof. As a result, no longer will there be a significant penalty associated with shopping in a global rather than a local or national marketplace. Rather, software agents will take care of irritating complexities like exchange rates, language translation, time scheduling, contracting, and project monitoring.

Both fuelling and anticipating such advances, a widely-backed spin off from Apple Computer known as General Magic is creating an open software agent middleware language known as *Telescript*.[33] This is already in use within AT&T's 'Personalink' agent-based communications system, and allows autonomous, communicating agents to share generic computer systems in a secure fashion. Such an ability is critical for successful software agent application. *Telescript* may therefore constitute a linchpin of technological and organizational convergence and future networked value creation.

Today, individuals may safely use the ATM cashpoints of banks other than their own thanks to agreements and systems for ubiquitous data sharing. In a similar fashion, *Telescript* and other comparable middleware will permit software agents to engage in digital business activities across computer network infrastructures which their masters do not own, and of which they are not even direct customers.

TOWARDS A WIRED MINDSET

Software agents, and indeed most other emerging middleware techno-logies, are intended to make networks smarter about people, rather than requiring people to become smarter about networks.[34] However, they are equally destined to remove economic and organizational complexity from our lives. Hence, when software agents capable of frictionlessly surfing the globe become the tools of a great many businesses and their customers, so a common and open technological *and* economic value-creating infrastructure will have arrived.

Today, interlinked networks and middleware advancements are increasingly oiling the cogs of the Wired Age. In turn, isolated islands of either technological or organizational infrastructure are finding it harder and harder to survive. For the business person, today's climate of connectionism and convergence poses an incredible challenge. Yet it is one that cannot be ignored. As Stewart Brand, biographer of the Media Lab at MIT, so nicely puts it, 'once a new technology rolls over you, if you're not part of the steamroller you're part of the road'.[35]

In order to prosper and create value within our emerging, networked landscape of technological and organizational convergence, many people need to evolve a new mindset. This *new way of thinking* will demand that most managers seek a range of technology competencies, whilst most technologists foster business skills. The markets of the future—and in particular the workers and customers of the *Windows* and Nintendo generation—will not understand or tolerate a business community in which 'computing' and 'business', or 'technology' and 'organization', are perceived as separate disciplines or career domains.

As Charles Wang, CEO of Computer Associates, argues in his book *Techno Vision*,[36] information technology no longer supports business. Instead, information technology *is the business*—with the systems of advanced, interconnected technologies and modern business organizations becoming inseparable. Indeed as Wang claims:

> So deeply embedded in every process that it is all but invisible, information technology will underlie every business activity in the connected organiz-ation. Information technology will not be considered a tool or even a way to leverage human knowledge. It will become an expected utility, much like electricity, noticeable only in the rare event it is withdrawn.[37]

A GLOBAL, ELECTRONIC MARKETPLACE

Spearheaded by the Internet, intranets and extranets, an interlinked global hardware platform has begun to emerge.[38] This will soon serve many technological *and* organizational functions. Indeed, many human communications, data exchanges, currency dealings, and media products, are already flowing freely across the global hardware platform's digital nexus.

As noted in the last chapter, technological hardware constitutes all of the physical elements of any system upon which software programs or other media content are executed. In a similar vein, organizational hardware equates to those structures, job roles, procedures, and regulations across which productive processes come to be overlaid. As businesses 'get wired', so many of the latter are coming to rely upon advanced technology systems. The economy of the future is the Digital Economy; its productive plant and trading conduits the computers and networks now ubiquitous within our offices, homes, and battery-heavy pockets.

With the mass network infrastructure of the global hardware platform already constituting a common technological *and* organizational vehicle for value creation, the whole nature of business activity is being transformed. In the past, world trade was focused almost exclusively on the exchange of physical goods. It was therefore both reliant on, and constrained by, available outlets, routes and technologies for product presentation and transportation.

However, as networks advance and interlink, an increasing proportion of trade is dependent upon the exchange of non-physical digital codes, rather than 'atom-based products'.[39] The stock in trade of a great many business sectors is thereby 'becoming bits'.[40] As a result, the trading constraints and opportunities facing many organizations now depend upon the availability of information infrastructures, rather than on the traditional business links of road, rail, sea or air.

In partial response to the above, a recent UK Government Technology Foresight initiative highlights how a new model of 'network centric computing' (NCC) is emerging. This signals radical changes, not just in telecommunications hardware, but in the business use thereof. As reported by IBM UK's Director of Technology, the network centric computing model attempts to more accurately jigsaw technology provision with the commercial needs of a modern enterprise.[41] It therefore reflects the evolution of business infrastructures in the light of increasingly-digital value creation.

Network centric computing also reflects a shift in business practice away from the defence of rigid organizational boundaries, and towards the management of value-adding processes. Increasingly, we are moving towards an economy based not on firms, but on networks thereof. In turn, many organizational networks are now both a function and a consequence of global, technological interconnection.

As often highlighted by exponents of task-focused organizational change, business process reengineering (BPR) and new information technologies (IT) today share a 'recursive relationship'.[42] In other words, the hardware and software of each has become vital in any evaluation of the other. Indeed, in espousing this 'new industrial engineering', BPR gurus Thomas Davenport and James Short suggest that:

> Thinking about information technology should be in terms of how it supports new or redesigned business processes, rather than business functions or organizational entities. And business processes and process improvements should be considered in terms of the capabilities information technologies can provide.[43]

As new business architectures and new technological infrastructures intermingle, new avenues for business action and interaction will continue to emerge. As Microsoft CEO Bill Gates argues, the 'information highway' is a bad metaphor for mass digital convergence and interconnection as it infers geography and travelling.[44] Instead, Gates suggests it is more sensible to think of the global hardware platform as a trading medium—or and as *the* market in which Adam Smith's vision of 'frictionless capitalism' may finally become a reality.[45] Indeed, according to Gates, the convergence of technology and organization will create for all a 'low-overhead shopper's heaven'.[46]

* * *

NETWORKS, MIDDLEWARE & VALUE CREATION

Across this chapter, we have begun to examine how networks and new middleware are transforming both technological valueware and business value creation. If such an exploration is to teach us anything, then it has to be that we need to remove in our minds the distinction between networks of computers and networks of organizations. Increasingly, the two will become indistinguishable. Indeed, in many instances, they have already come to rely upon the same hardware and software in the transfer of value-enhancing inputs and outputs.

The success or failure of many a future computer network or networked organization is likely to depend upon two main considerations. The first will be its architecture, and the second its user interface. However, both may increasingly be addressed from a single technological/organizational perspective.

Several modern organizational models are based around a central core upon which outsourced partners and customers draw.[47] The similarity of this topology with the 'client–server' (or 'client–host') model of modern computer network architecture is hopefully immediately obvious. Both feature a sharable, control resource at their notional centre. Both also do so in order that a changing variety of autonomous clients (or 'agents') may connect in to add value to themselves, and/or to the core/host resource, as and when required.

As long known by researchers into effective human–computer interaction (HCI), the interface that enables a human being to communicate with any technology is critical if the overall system is to prove a success. Good interfaces engage users in fluid, satisfying interactions that feel natural and well choreographed. On the other hand, bad interfaces inhibit effective user–system interaction. They can therefore lead to poor user perceptions of the system's value-enhancing potential as a whole.[48]

As electronic consumer–organization interaction becomes more common, so more and more people are going to *experience* organizations entirely through software code. Many remote workers will also engage in the majority of their interactions with organizations via the software interface of a network link. In order to build effective customer, worker and partner relationships, many companies therefore need to learn from the discipline of HCI.

Indeed, the more widely organizations use networks and middleware such as the crossware of the Internet, the more likely many people are to perceive and treat companies as if they were no different from any other desktop PC application. When and if this proves to be the case, organizations may be forced into the software industry's game of regular interface updates cascaded with ever more new features, and topped-off with the latest populist look and feel. If this happens, the valueware concept will in turn really be pushed to the fore.

In their attempts to cope with such looming technological and organizational convergence, companies and their managers quickly need to heal any 'disconnects' that may exist between their IT function and the rest of the organization. Such rifts are common, increasingly well documented, and often culturally founded. In particular, 'business managers' are often reported as not appreciating the role that IT can play in business transformation. Conversely, IT managers are often criticized for being too focused on technology and its development, rather than on customer and user requirements, and/or cost-effective value creation.[49]

In order to heal such an increasingly critical divide, it is widely acknowledged that senior managers need to take a much more active interest in IT and its application. Several analysts have also suggested that 'hybrid managers' ought to be employed, with both technological and business knowledge.[50] These individuals may then be charged with mediating between an organization's business and IT camps in order that a maximum technology/organization valueware overlap may emerge.

As the concept of doing 'cyber business' transmutes from fantasy to reality in many minds as well as across many organizations, so economic value creation will increasingly become a digitally-mediated and even digitally-delivered phenomenon. Our emerging global hardware platform of interconnected networks will in parallel rise as our planet's primary trading gateway. Advanced software agents may also become common 'virtual workers' that will remove many business frictions from our lives on a great many levels.

In light of the above, many people may reasonably start to wonder how far the convergence of technological and organizational hardware and software may go. Will 'networking' and 'business' become terms almost meaningless in isolation? Are tomorrow's managers destined to become as much 'programmers' of their organizations as they are strategists and leaders of women and men? Indeed, to what degree are the very *identities* of both ourselves and our organizations likely to be altered in a new world in which so many traditional boundaries look set to crumble?

The above questions are broad, complex, and open to wide philosophical as well as practical interpretation. However, they provide substantial mental fodder for chapters to come. Not least they lead us nicely into **chapter 3**, which sets out to examine that increasing conflict between 'flexibility' and individual 'identity' that a great many people are now starting to face.

3
Flexibility
or Identity?

JUST LIKE TODAY'S COMPUTING infrastructures or organizations, most individuals are networked entities. Indeed, the complexity of the modern world increasingly forces people into ever more intricate patterns of fluctuating relationships.

Touched by and constantly touching others, we are built and weathered by every human contact. Our very sense of 'self' is derived from our perception of affiliation and contribution. There can be no such thing as 'one' without a wider 'many' to reflect us back.

The more solid our network of relationships, the more secure our sense of identity. It is therefore hardly surprising that many people have been troubled by the recent drive for increased individual and organizational flexibility.

Advocates of the flexible workplace argue that its fluid make-up is now essential if businesses are to respond as rapidly and cost-effectively as possible to changing market conditions. Time and time again, we therefore hear the cry that flexibility has become a prerequisite for the survival or advancement of many companies. Indeed, **chapter 5**'s summary of contemporary management approaches clearly highlights the perceived importance of flexibility in achieving corporate success.

This chapter does happen to be all about ever-increasing flexible and 'free agent' working patterns. However, it remains free of the usual pro- or anti-flexibility doctrines. Rather, the purpose of the following is to examine the impact of fluid and uncertain workstyles upon individuals as both workers and customers.

Consideration will also be given to the influence of harsh new economic and technological realities upon the blurring boundaries between humanity and organization—or, more specifically, between our social and working selves. As a result of these two threads of study, it is hoped that a broader understanding of the post-economic *impact* of flexible working on value and its creation will emerge.

WHO DO YOU THINK YOU ARE?

'Who are you?' is a question of fundamental importance to most individuals. It is also a query that, with some thought, most people may answer in a great many ways. Instinctively, however, when asked who or what they are, most men will reply with their job title, and often the name of their employer. At least nine times out of ten, the label of 'architect', 'computer programmer', 'manager', 'salesman', 'administrator', 'builder', 'plumber' or whatever is self-attributed before that of 'husband', 'father', 'son', 'gardener', 'football supporter', or any other dominant social tie or life activity.

Women remain more likely than men instinctively to cite a family or social role before a work-related position when asked who or what they are. However, even when not voiced at the top of the list, work affiliation is still highly valued when it comes to defining a woman's actual or desired sense of identity. The retired also often describe themselves with reference to their former employer and occupation. And many school children and students still aspire towards not just *any* job, but instead a specific occupation that will in time provide them with a security label.

There is, of course, nothing intrinsically wrong with the above state of affairs. It may well be that total personal happiness is unlikely if how one instinctively describes oneself, and what one *values*, are not in harmony. This said, few people have the time, energy or opportunity to match these two measures. We may all dream. However, most of the time, providing that we can securely define ourselves as somebody—as *anybody*—then a reasonable degree of rutted contentment is likely to result.

Unfortunately, rutted contentment is often the victim of the new religion of flexibility. For many, past linchpins of stability are being swept aside in the name of economic necessity, and with the support of emerging technological tools that allow us to work in new ways. Today, literally millions of people across thousands of organizations are being forced to flexibly adapt into a harsh new world whose comforts are as yet largely unknown. And as this happens, the casualties in human terms are invariably high.

OUR ERODING ANCHORS OF SELF

Today, the drive for flexibility in the workplace is threatening a great many people's sense of self in two distinct ways. Most obviously, economic pressures for flexibility are weakening or removing the certainty of previously secure and long-term worker–organization relationships. However, more subtly, changes in business hardware—in the physical and communications infrastructures used by organizations—are also robbing people of clear distinctions between their 'work' and 'home' lives. For example, most teleworkers tend to find that their domestic activities become highly interwoven with periods at the keyboard.

As an old saying reminds us, all work and no play is likely to make Jack or Jill rather dull indeed. Most of us desire a distinct 'working life' and a separate 'home life' in order to remain rounded individuals. The overworked hence dream of holidays, whilst the unemployed usually desire a job. We instinctively *feel* when something is missing. We have come to be defined by two or more 'boxes of self', and in particular by the tensions and/or interdependencies between them.

Figure 3.1 illustrates the foundations of many people's sense of identity as stemming from a combined 'work' and 'home' duality. Consider it for a minute or two. Just how would you categorize these two boxes in your own life? Are they clearly distinct, both in balance, and equally fulfilling? Or is one 'box of self' largely lacking and/or dominating the other? What are the key tensions and interdependencies between your own 'work self' and 'home self'? Are you a workaholic in need of a (home) life? Or do you loathe work and endure it purely to support other activities and pleasures?

Box 3.1 contains a list of often-desired rewards and qualities towards which many people aspire. Consider for a minute or two how many you can personally attribute to the experiences and rewards you derive from your work and home lives. How many may you attribute to both? And

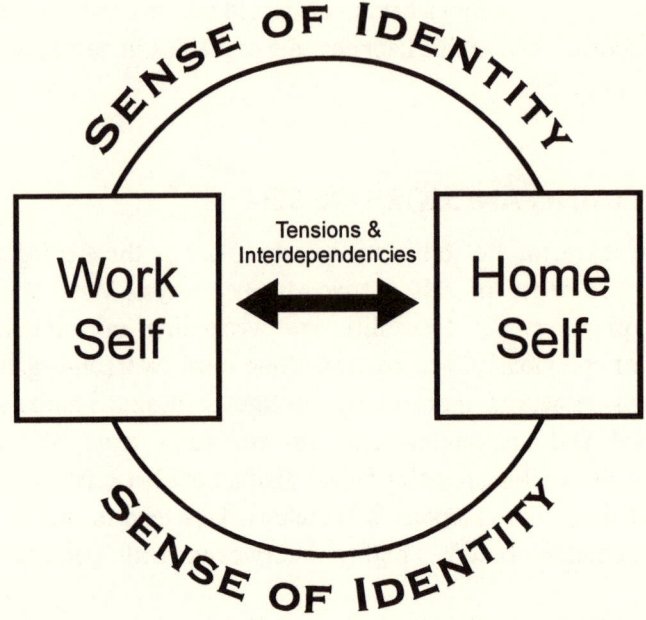

Figure 3.1 Identity as a Function of Two 'Boxes of Self'

how many rewards and qualities attributed to one box depend upon the activities and/or experiences of the other?

Try adding some more rewards and qualities to the list, and then attributing these to your 'work' and/or 'home' boxes. This is sometimes a difficult exercise, though one whose results are often surprising, and to which we will return in later sections.

NEW WORKING RELATIONSHIPS

Across history, the means by which individuals have rendered their labours to wider organizations have radically evolved. For thousands of years, many human beings were explicitly or implicitly owned as serfs or slaves. Then, as industrialization dawned and capitalism demanded a paid

---

□ Physical sustenance and shelter
□ Social interaction
□ Monetary reward
□ A sense of belonging
□ A sense of achievement
□ Excitement
□ Stability
□ Security
□ An escape
□ A familiar routine

Box 3.1 Rewards and Qualities of our Work and Home Lives

workforce to purchase its wares, the dominant work/command relationship altered to that of employer with employee.[51]

Such a largely long-term and secure state of affairs dominated the working lives of most Western individuals until around the mid-1980s. However, since that time, recession-driven global pressures on labour costs, coupled with an accompanying decline in trade union power, have led to the erosion of this previous 20th century norm. As a result, a growing number of people now have implicitly or explicitly to trade themselves as 'free agents' within harsh and insecure open labour markets.

The emergence of a growing 'secondary' or 'peripheral' workforce was highlighted by John Atkinson and the Institute for Manpower Studies in 1985.[52] Their analysis suggested that an increasing number of firms were only taking on new staff on a temporary, part-time, fixed-term, or even zero-hours basis.[53] The proportion of people working as 'core' employees within organizations—and hence enjoying the security and non-pay benefits of such employment—was therefore starting to decrease.

Not much more than a decade later, a 'labour periphery' workstyle seems to be becoming the norm. Management guru Charles Handy has for several years calculated that around fifty per cent of the potential

Western workforce no longer enjoys a secure, traditional employment relationship.[54] Similarly, a report from the Institute of Management and Manpower Plc recently predicted that, by the year 2000, between twenty-five and fifty per cent of the workforce of most major organizations will be comprised of temporary and contract staff.[55]

Even the above figures only reveal part of the trend towards workplace free agency. For a start, added to those workers who only have short-term relationships *with* organizations, there are a growing number of people who have no organizational parent whatsoever. In the United Kingdom, for example, there are now around 3.3 million people registered as self-employed, representing some fourteen per cent of the total labourforce.[56] In addition, with the rise of teleworking, even those still in long-term employment are starting to experience far less specific workplace identities.

In the United States there are now around nine million telecommuters, with eighty per cent of companies having adopted teleworking in one form or another.[57] The ranks of such home-tied individuals are also growing at fifteen per cent per annum. Across the Atlantic, there are some 560,000 telecommuting workstations in the United Kingdom,[58] whilst the European Union has a target for ten million teleworkers across the community by the turn of the millennium.[59]

What all of the above figures imply is that a larger and larger number of people must now thrive or fight to survive as free agents upon the edges of traditional organizations. However one views the morality of this situation, from a business perspective the reason is simply that free agent, periphery workers impose far fewer labour obligations (and hence costs) than overhead-hungry core employees.

Even within today's lean-and-economically-mean companies, most core workers continue to reap benefits ranging from holiday, sickness and redundancy pay, to pension contributions, and perhaps even personal health insurance. A significant proportion of core workers also enjoy other job-related perks such as company cars.

In comparison, most periphery workers receive no such rewards. Instead, whether defined as 'self-employed', or else surfing in and out of temporary or contract work as and when available, they survive as what Handy has termed 'portfolio people'.[60] Indeed, in terms of labour economies, each and every worker orphaned from a secure, employment guardian effectively becomes an outsourced organization of one.

FREE AGENT LIVES

Often starved on choice, 'free agent' workers are only free in the sense that with few organizational obligations owed to them, they in turn owe few obligations back *to* particular, single organizations. Indeed, in terms of basic personal economics, many workers simply can't afford to foster one-to-one company ties.

As a result, far more people than ever before intermix self-employment, multiple part-time jobs, short-term contracts, and periods of temporary/agency work.[61] In fact, many of those who sell their services freelance to others—be they management consultants, trainers, caterers, media creatives, computer programmers, gardeners, builders, financial advisors or whatever—can only achieve any kind of stability in their working lives by building up a portfolio of clients, the loss of any one of whom will not prove critical.

Many free agent workers also rarely have fixed hours to start work because they rarely truly stop. Instead, their 'work' and 'home' boxes of self mingle into a portfolio life that may have a myriad of elements, but from which it is hard to find any escape. In particular, for free agent knowledge workers, computer networks and new middleware technologies are increasing the pressure to become a constantly-available resource.

Unlike paper-and-envelope snail mail, e-mail can be waiting to arrive twenty-four hours a day. This is especially true for those who trade themselves digitally in a global market. Multiple, interactive, multimedia document exchanges can now occur in the space of hours or even minutes. Consequently, there is no longer a respite granted before client feedback after something has been dropped in the post.

As discussed in the last chapter, networks and new middleware are rapidly driving electronic business disintermediation and are forcing knowledge exchanges into real time. Going on-line and responding to an e-mail at 1:00am in the United Kingdom (perhaps mid-afternoon for a prospective client in the United States), can for many already mean obtaining work that would no longer be available at 9:00am the next morning.

Customer service on a broad range of levels is now also often demanded around the clock. As players in a market economy with increasing 24-hour demands, we therefore all have no reasonable right to be surprised that so many offices and an increasing proportion of individuals can rarely put up the shutters for the night. Even those in the traditional labour core are subsequently being forced into the free agent mentality of becoming a constantly on-line human facility. In short, as the third millennium dawns, fewer and fewer of those in work will be able to boast clear edges in terms of their time availability or physical location. Defining the 'what', 'when' and 'where' of our working-self box will therefore become more and more difficult.

THE VALUEWARE CHALLENGE

For far too long, coping with new technology has alone been headlined as the mightiest challenge of the future. However, this is only because few yet see (or want to see) that the greater barriers to value creation ahead lie in understanding and adapting to today's social and cultural evolution of who we are and how we work.

Free agency (as a state of mind as well as a practicality) is a complex issue whose broad implications few managers or organizations yet understand. Most probably, this is because the aftermath of more and more flexible and free-agent working practices will impact upon many managers and organizations in two distinct ways.

Companies that continue to depend upon an increasing proportion of flexibly-contracted workers must eventually face the need to begin managing and motivating their labourforce in new ways. However, the implications of an enlarged free agent labour pool will not end with new human resource management practices alone. This is because there is a special word for any large number of workers. And this critical collective noun is 'customers'.

Those people who no longer enjoy clear, organization-specific security and identity will, in aggregation, transform the nature of many consumer markets. To cite just one simple example, individuals facing an insecure workstyle will be less likely to take on long-term financial commitments than their traditional employee cousins. As a result, as noted in a recent bulletin concerning the implications of increasing self-employment:

> Banks, financial advisors and public support agencies will have to develop a better understanding of what it means to be self-employed. They will have to train their staff to develop empathy based on human factors, to help reduce the feeling of isolation, [and] to build confidence by reducing anxiety.[62]

As the above begins to suggest, in order to fulfil human needs no longer met by an employing parent organization, free agents as customers may begin to demand new affiliations (as well as more flexible services) from the companies with whom they choose to spend their hard earned cash. The economic and organizational impact of an increasingly flexible, free agent *society* will consequently lead to both new worker/organization *and* new customer/organization relationships, as illustrated in **figure 3.2**.

MANAGING FREE AGENT WORKERS

As the percentage of their workforce employed within their core continues to diminish, so those organizations seeking to extract maximum value from their labour inputs will need to reassess their human resource management strategies. Unlike traditional employees, free agent workers do not have a career ladder before them as a motivator towards optimum performance. Nor do they have a long-term status, a company car, a pension, or a health care plan, to lose as a result of switching to another work activity.

In response to the above, some have argued that we are moving towards a more 'perfect' labour market in which the only motivator is money. Indeed, on occasions I have heard managers preach that people with part-time, temporary, short-term, or even zero-hours contracts, ought to feel 'lucky to have any work opportunity'. Yet, even in the short-term, such a point of view has to be naïve.

Money is clearly vital for all of us seeking to survive and thrive in the bitterly-harsh and welfare-provision-poor market economies of both today

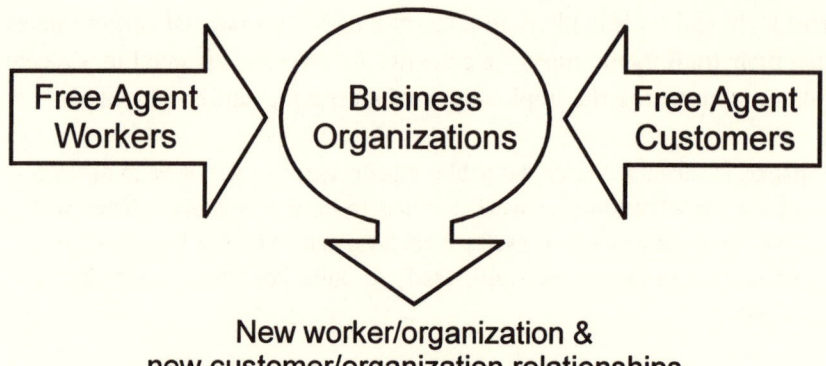

Figure 3.2 The Dual Impact of Flexible Labour Practices

and tomorrow. However, this point firmly noted, very few people are actually motivated to *excel* by money alone. It may well be that it is a financial reward (and/or the risk of losing the same) that keeps most people labouring throughout their working night or day. Yet, as many studies have shown, it is nevertheless non-financial rewards that usually play the greatest role in determining whether most people do a really good job, or merely an adequate one.

A sense of being part of a cohesive working *team* or *community* or *culture* is an important motivator for the majority of human beings. Although most people moan about their work, deep down they usually also value the way in which it makes them more than an individual. As noted at the start of this chapter, over time a strong working role and routine bestow both identity and the incredibly powerful motivator of self-worth.

Today, almost by definition, most free agent workers (including many teleworkers) remain unlikely to experience work as a shared, communal and self-enhancing undertaking. Just look back at the list of rewards and qualities that individuals may gain from their work and home lives in **box 3.1**. How many of these do you believe are likely to be earned by factory- or office-present, long-term employees alone?

This question is clearly a matter upon which many managers need to reflect if they are to get the best out of free agent staff. Many companies continue to exclude temporary or sub-contracted workers from employee perks ranging from cut-price offers, to the membership of social clubs, the use of sports facilities, and sometimes even free or subsidised refreshments. Such restrictions that divide core staff from the labour periphery undoubtedly save money. Yet they may cost more by removing the motivation of many free agent individuals to labour for job *quality* rather than *acceptability*.

The economic-essential business paybacks that spring from flexible employment practices are not going to go away. However, there remains a fine line to be drawn between forcing an increasing proportion of a company's labourforce to work solely for money, and the provision of enough non-pay motivators to maximize its aggregate contribution.

A simple rule to remember is that the greater the variety of value paybacks an individual may obtain through work, the more likely they are to contribute a maximum level of effort. And this is almost irrespective of how much they are paid for their time in terms of money.

As noted earlier in this chapter, people tend to feel more secure and content in themselves the better the balance between their 'work' and 'home' lives. It is therefore socially unfortunate that, as a new millennium dawns, the logics of economics and flexibility are conspiring to reduce the social and motivational elements of many working environments.

The result of the above is a wild unbalancing of many individual's 'work' and 'home' boxes of self. Managers as leaders of women and men would therefore be wise to take any steps possible to hold the tilting scales level. In particular, managers may like continually to question why anybody should want to work for their organization, and why too their workforce should strive to deliver their best possible service to its customers.

THE FREE AGENT CUSTOMER OPPORTUNITY

As purchasers of human time, the challenge before many Future Organizations will be to reward their workers with as rich a sense of achievement, belonging and identity as possible. Unfortunately, whatever the hopes of the previous section, this remains a quest in which many companies will largely fail. However, as a consequence, many Future Organizations as marketeers will be presented with a remarkable opportunity. And this will be to package and sell back to free agents as *customers* those valued feelings of security, kinship and self-esteem that flexible workstyles continue to take away.

Across the 20th century, most trade has been in tangible goods and short-term, physically-manipulative services. Intangible, long-term group affiliations have rarely commanded a high financial price. However, this has only been because extended families, religions, lifetime organizational parents, warring nations, and strong local communities, have generally provided the latter for free.

Perhaps unfortunately, few of the aforementioned traditional mainstays of a healthy society appear destined to continue to dominate all of our lives. We have already noted how many individuals are being robbed of a secure, organizational life-anchor by increased workplace flexibility. However, the legacy and impact of our increasingly individually-centred, free agent society also runs far deeper into our collective soul.

Just as new workstyles now enforce a 'me-first' mentality, so growing numbers of the young and old, families and single people, now lock themselves away from a world of perceived insecurity. The fact that our 'unsafe' planet is almost completely populated by millions of other equally nervous human beings is largely ignored. As a consequence, our natural, concrete environment inevitably offers fewer and fewer social spaces within which to nurture new friendships and community affiliations.

Poverty, rising crime, drug cultures, and a younger generation disenfranchised from employment, may seed this sorry spiral. However, these oft-blamed, dark-street ills are probably secondary causes of human isolation—easy targets to blame when the real roots of our troubles lie in the fact that most of us cherish too dearly the social-free-agent sanctity of locking our doors.

Across most prosperous nations, there is a growing realization that many workstyles and homestyles boast more than enough goods and services, but far too few lasting community ties. A clear market opportunity is therefore emerging to sell to many people that sense of community and group affiliation that their parents and grandparents once took for granted. Those companies that recognise this, and that learn how to sell groups of people back to each other, may hence become the corporate giants of the 21st century.

NEW MARKET REALITIES?

To some, highlighting the potential sale of community relationships as a 'market opportunity' may appear sad, cold, mercenary, and perhaps even exploitative. However, the fact remains that capitalist markets have become the dominant means for meeting human needs and solving human problems. What's more, this has only come about because most of us have spent and wished us all into championing this now-dominant reality. To complain that we should not in future seek long-term warmth, certainty and kinship through the cold mechanics of markets and their technologies therefore has to be somewhat hypocritical.

One might even contend that, by turning community affiliation into a market commodity, we may enable a gentler, more caring mode of capitalism. Indeed, the greater the number of firms that decide to trade in group identities, the greater the market supply will be. In turn, the lower prices will fall, and the more broadly 'affiliation products' will be able to be afforded by the majority.

Like the air around us, we've taken our group identities and relationships for granted for so long that we have begun to value far too exclusively those goods and short-term services in which we money-trade. In other words, it may be that we need to review what we have chosen *to* value, rather than the mechanisms by which we *trade* value. We would be foolish to ignore the possibility that capitalism may be able to evolve to sell us an environmentally-friendly social atmosphere. Certainly, you don't have to knock on many locked doors to find people who would prefer to pay for stronger community ties, rather than more polished metal, throwaway plastic, or *Windows N+1*.

MARKETS IN AFFILIATION

It may prove difficult to visualise markets that sell their customers that sense of kinship, community and 'wholeness' that modern workstyles and even homestyles no longer supply. After all, the tangible exchange of physical goods has near-dominated human trade from the reign of the early, Near Eastern empires. Even since that time, many of those services described as 'intangible' by schoolbook economists have nevertheless continued to render physical results—be they in the form of cleaned floors, serviced machines, decorated buildings, served meals, cut hair, transported goods, relocated human beings, or debugged computers.

To date at least, a far narrower range of services has rendered purely mental results. Education and training may at first appear to have no tangible output. However, they usually result in the ability to make or alter a physical object, or else more expertly to undertake some physical action.

Today, almost certainly the majority of business organizations to successfully market human emotion or mass affiliation are to be found within the industries of media and entertainment.[63] Other companies wishing to discover new affiliation products to sell to free agent individuals may therefore be wise to turn their attention to some existing worlds of collective fantasy.

THE STAR TREK PHENOMENON

It is over thirty years since the *Starship Enterprise* departed on its first mission to boldly go where no one had gone before. Since then, the original *Star Trek* shows—together with three sister series and eight feature films to date—have become not just a science fiction cult, but also the foundation of a highly-cohesive global tribe. There can be little doubt that, back in 1966, producer Gene Rodenberry had little idea just what a powerful force for human kinship he was creating.

A great many people know a 'trekkie' or two. Some (and we are talking here of approaching forty per cent of the Western viewing population) simply watch the TV shows. Others additionally buy books and videos, dress as their favourite characters, attend numerous conventions, and even learn the science of *Star Trek* technology or the vocabulary of the 'Klingon' language.

Many non-fans consider such an 'extreme' devotion to 'just a television show' to be mad or sad or both. Most commonly, cynics contend that trekkies have simply 'not grown up'. However, those who make such a judgement are perhaps missing the point. People buy in to *Star Trek* (or the equally-global science fiction tribes of *Star Wars*, *Doctor Who* or *Babylon 5*) not to maintain a link to their childhood, but to become part of an ongoing fan community with an overarching mythos into which they can always escape.

In a world of fewer long-term organizational parents or extended families, and in which the traditional, tired institutions of nations or religions often hold little attraction, fictional 'controlled realities' have for many become powerful life-anchors. Today this is reflected in the exponential growth of science-fiction related product ranges targeted at those in their twenties and thirties, as well as at toddlers and teenagers. Granted, in the past firms such as Disney successfully re-released classic fantasy films to generation after generation. However, the difference today is that fantasy gurus (such as *Stars Wars* creator George Lucas) are managing to maintain the simultaneous devotion of older generations to their artworks *in addition* to winning fresh converts. In comparison, nations and many religions in the Western World are struggling hard (or not at all) to capture young hearts and minds into those group loyalties to countries or gods so powerfully exhibited by their grandparents.

As the above single case hopefully highlights, some business organizations are already becoming significant sellers of affiliation and community. Drawing together integrated multiple media—from films and videos to books, models, theme parks, role playing games, costumes and CD-ROMs—they are creating long-term 'metamedia' experiences as security anchors to help lessen the stresses of free agent lifestyles.

Creating global brands that support the membership of global tribes will be one key to success in the relationship-rich markets of the 21st century. What's more, the tip of this new market iceberg is not only visible in the fictional worlds of outer space. Already Coca-Cola, McDonalds *et al* are using metamedia marketing to export global cultures to consumers who in turn demand global brands that permit them to become part of market-cultured global tribes. Everyone, everywhere has a need to belong to something large enough to make them feel secure, yet small enough to allow them to be recognized as different.

Like it or not, advanced transport systems and computer networks are turning more and more of us into global citizens. Unfortunately, such a planetary affiliation provides no more identity than being a member of the human race. As noted at the very start of this chapter, there can be no such thing as 'one' without a wider 'many' to reflect us back.

In comparison to those who still harbour significant national affiliations, global citizens have no wider 'many' against which to identity themselves as a sub-set. They may therefore discover a growing need to buy in to something with a broad but limited boundary. Once this might have been an organization. And it might be so again. The difference is, the organizations that individuals will build life affiliations with in the future won't be those they *work for*. They'll be those they *purchase from*.

BUILDING TRIBES IN CYBERSPACE

For several years, businesses have been searching for a means of generating significant value from the Internet. Today's billboard web sites and on-line shop fronts may be all very well when it comes to advertising, supporting, and perhaps even *trading* traditional products and services. However, few companies have actually begun to make significant financial returns from an Internet presence alone.

The above, however, may be about to change with the emergence of on-line 'virtual communities'. By combining niche-market content with the many-to-many communications model of the Internet, these provide social forums with which members may develop lasting affiliations. Typically the point-and-click interface of the world-wide web is used to make such forums easy to access, as well as to provide them with a familiar, branded look-and-feel.

The companies that host virtual communities usually 'seed' them with a variety of discussion groups or 'conferences' of potential interest to their target membership. Each conference is then divided into 'topics' to which members may contribute text-based messages or 'posts'. A range of graphical controls allows members to interact with conferences, topics and posts at various levels, as well as to customize a personal interface. Personal web pages are often also available to members, as are real-time chat rooms, archive material, and trading zones in which to make or discuss on-line purchases.[64]

According to influential Harvard gurus John Hagel and Arthur G. Armstrong, virtual communities provide an emerging economic model for meeting individual relationship, interest, fantasy and transaction needs.[65] So the argument goes, virtual communities enable people to connect in on-going relationships regardless of the traditional constraints of time or physical location.

From a business perspective, virtual communities also have the potential to become key future 'infomediaries' that will champion member interests by shifting market power from vendors to consumers. As Hagel and Armstrong contend, this will happen as the 'range, richness, reliability and timeliness' of the product information interactively exchanged within virtual communities becomes greater than that in any other market forum.[66]

As suggested by Mark McDonough, a pioneer of International Thomson's virtual community labs, on-line virtual communities may generate direct subscription revenue, enhance a brand/market position, provide valuable customer feedbacks, and create new market opportunities.[67] In fact, a successful on-line virtual community ought to define its parent business as *the place to go* for contacts, credible information, and relevant content. This may particularly be the case if the virtual community serves a narrowly-defined but geographically-distributed member base.

Virtual communities will prove of growing importance as the worldwide web becomes the default interface for on-line, desktop communication. Creating and nurturing successful virtual communities may hence become a very big business. This is because those communities that forge strong enough member affiliations to ensure habitual, daily visits are likely to become many people's 'front-ends' to the on-line, digital marketplaces and social spaces of the future.

In effect, by making other customers part of the product, on-line virtual communities extend customer-organization interaction beyond relationship marketing. In the near future, virtual communities will become cyberplaces to capture and share the *experiences* of thousands of real people who will value not just the brand of the supplying organization, but also the certainly and friendship of other *specific* individuals being on-line.

The implication of the above is that the virtual community business will rapidly become one of increasing returns. Fairly obviously, the attractiveness of any virtual community will increase the greater its number of members and the volume and quality of its on-line content. The big will therefore get not just bigger but more profitable, as early entrants to the virtual community marketplace aggregate critical mass and member loyalties.[68] Hagel and Armstrong's claim that 'a low barrier to entry business is almost certain to build insurmountable barriers to entry over the next five years'[69] is therefore hardly surprising.

It is, perhaps, too easy to get sucked into the evangelism of those on-line pioneers who claim that virtual communities will spawn 'new, rarely capital intensive industries' for 'souls in search of relationships' as well as 'customers in search of products' and 'suppliers in search of markets'.[70] However, free agent workers seeking anchors of certainty in their lives must be extremely likely candidates to affiliate to such easy-to-access social and business forums.

Of course, this need not imply that every organization under the Sun should go out and build its own on-line virtual community. However, what it does imply is that those companies who wish to remain in business ought at least to *consider* the creation of their own, interactive on-line tribe. Or, at the very least, most businesses would be wise to devise a strategy to *inhabit* at least some of those virtual communities likely to be frequented by their existing and future customer base.[71]

* * *

MAINTAINING LIFEPLACE TIES

In his fascinating book *Connexity*, Geoff Mulgan highlights the paradox of how mass individualism breeds mass interdependence.[72] Across the world, just when the boundaries of nation states are collapsing, organizations no longer demand our long-term devotion, and the tools of the Digital Age are permitting us to work any time and anywhere, so the fact that most of us remain incapable of survival without the support of the majority is finally hitting home.

Human beings are social animals. Whatever some may claim, most of us remain poor at surviving in isolation with any degree of emotional comfort. Only individuals bonded to long-term human groupings may possess those strongly desired, stable identities that known others may reflect back. As Peter Davies, Managing Director of Prudential Insurance, recently pondered in one of his company's television advertisements, 'even adults need a security blanket' to bring a sense of long-term safety and even *meaning* to their lives.

Unfortunately, as we have explored across this chapter, the security blanket of a long-term job is under threat, whilst for many the ability to 'escape' from home to work (or vice versa) also continues to erode. For many individuals, the challenge of the future will be to adjust to a lifestyle with fewer discrete, long-term boundaries. For Future Organizations, it will be to learn the new preferences of those free agents who, as consumers, no longer enjoy two (or more) clear components of personal identity.

By putting a price on individual affiliations to crafted causes or cults, it may be possible for some Future Organizations both to thrive *and* to restore a more secure emotional balance for the next generation of free agent workers. One way they may do this is by using complex 'metamedia' to construct 'controlled reality' fantasy worlds to which individuals with little organizational identity will willingly surrender part of their lives. Alternatively, other companies may create on-line virtual communities to which free agents as customers may choose to affiliate in order to be provided with a long-term, social forum in addition to a digital shopping mall.

What the latter suggestion implies is that silicon, plastic and metal technology may have an increasing role to play in fostering real 'human' identities in a world of flexibility, free agency, and mingling workplace/homeplace boundaries. However, a level beyond, the growing interplay of humanity with its saviour or nemesis—technology—may soon fuel far more powerful value engines and convergence forces. Indeed, as we shall explore in the final chapter of **Part I**, the day may not be far away when we will find ourselves seamlessly as one with our machines.

4
Playing at God

ACROSS HISTORY HUMANITY HAS wrestled with the boundaries of its own identity. For thousands of years, human beings positioned themselves at the centre of the universe. Admittedly, some ancient Greeks questioned this separation of humankind from the rest of creation. However, it was not until 16th century astronomer Nicolaus Copernicus proposed his theory of heliocentricity—with the Earth rotating around the Sun and not the other way around—that the perception of Man and Woman in the Universe started to change. Over the following centuries, a realization subsequently took hold that humanity was just one part of creation, rather than its dominant epicentre.

When Charles Darwin popularised his theory of evolution, a second radical questioning of our race's uniqueness was triggered. To the alarm of many, Darwin proclaimed a new order in which human beings were members of the animal kingdom. Men and women were scientifically linked to all other animals through the survival of the fittest and sexual selection. Darwin was hence daring to dispute God and *Genesis* as the force behind our own creation.

Many Victorians were disturbed by evolutionary theory. No longer could human beings cling to a position of dominance over nature. Or as T.H. Huxley declared, Darwin's theories nullified the veneer of civilization as Man's 'escape from his place in the animal kingdom'.[73]

Yet a third questioning of our very nature arose from the work of Sigmund Freud. His genius was to forge a rational science out of the irrationality of language and dreams. As a result, Freud highlighted human beings as psychological as well as physiological creatures.

More alarmingly, Freud linked our psychology to our evolutionary, sexual nature. In doing so, he removed not just the previous separation of the conscious from the unconscious. In addition, he questioned the illusion that rational human beings dominated their irrational, unconscious selves. Of all of the brave revelations that have ever caused us to question our own nature, Freud's may therefore always remain the most disturbing.

THE FOURTH DISCONTINUITY

According to Professor Bruce Mazlish, when Copernicus removed humanity from the centre of the universe, Darwin linked us into the animal kingdom, or Freud placed our conscious and unconscious on a spectrum of interdependence, clear 'discontinuities' were relinquished by humankind.[74] Mazlish argues that human beings have always had a need to feel special or superior, and that even though we now perceive ourselves as part of nature, there still remain mental barriers for us to overcome.

In particular, Mazlish suggests that we are currently on the threshold of breaking past the 'Fourth Discontinuity' by which we define ourselves as being different from machines. Such a thesis is in two parts. Firstly, Mazlish states that it is no longer realistic to think of humans without machines. Secondly, he suggests that the same concepts now explain the very *workings* of both human beings and many artificial mechanisms, and that as such they bridge any sharp divide between the two. According to Mazlish, a continuous scale is therefore emerging with human beings at one extreme, machines at the other, and an increasing region of commonality in between.[75]

In the first chapter I suggested that, in an age of increasing convergence, the boundaries between humanity and technology are starting to blur. Within this context, Mazlish's thesis becomes an exceptionally powerful conceptual tool. Across this chapter, some of Mazlish's ideas will therefore be used as a springboard as I try and predict the hope or terror of future human valueware.

For decades if not centuries, many people have feared the human race's growing love affair with technology. Yet, if Bruce Mazlish is correct, it may yet prove a proactive marriage of synergistic interdependence that can only make us *more* rather than *less* 'human'.

BARRIERS IN OUR MINDS

For most people, the apparent differences between technology and humanity remain multitudinous. Humanity is alive and natural, whilst technology is dead and artificial. Human beings are able to think and to manipulate their environment, whilst technology is incapable of conscious reasoning or undirected transformative action. Human beings are biological entities with the ability to reproduce themselves, whilst technology has to be manufactured from inert materials like plastic and metal.

As the Fourth Discontinuity continues to erode, few of the above distinctions will continue to hold true. For example, we already live in a world in which *biotechnological* materials are fabricated and grown. Behind some corporate facades, software programs also mine vast databases in search of questions as well as answers that sometimes result in the manufacture or shipment of goods with no human intervention.[76] In addition, many people now depend for their survival, or at least for a decent quality of life, upon metal-and-plastic prosthesis.[77]

What the above indicates is that even today no longer are all technologies non-biological and passive, with only human beings capable of acting independently to manipulate the physical world, and only technological machines engineered and 'manufactured'. Separating humanity from technology in terms of biological versus non-biological, natural versus artificial, or designed and manufactured versus evolved, is hence increasingly problematic. As a result, one may debate at length the most appropriate ways in which 'humanity' and 'technology' may now be best scientifically or philosophically distinguished.

Such a discussion could draw from a wide range of learned literatures and opinions. However, at its close we would probably be no more enlightened in practical terms. We are probably therefore wisest to live with very simple, perceptual conclusions. As a result, and as already noted in **chapter 1**, herein I will define 'humanity' as that which we now *choose* to recognise as human, and 'technology' as the collective nature of all of those tools, programs and media that human beings use at work and at play. Further, as also discussed in the first chapter, both humanity and technology will be envisaged as possessing hardware and software components that, in amalgamation, contribute their respective valueware.

HARDWARE PUSH & SOFTWARE PULL

As highlighted by Bruce Mazlish, humans and machines will increasingly be conceptually distinct only when we choose to perceive them as such. However, this need in no way prevent the study of the convergence of humanity and technology—and of the valueware overlap so created between the two—from proceeding in a structured fashion.

As illustrated in **Figure 4.1**, two sets of broad hardware factors, and two sets of broad software factors, may potentially drive any further convergence of human beings with technological systems. Of these, the hardware factors comprise new human–computer interfaces, as well as those medical developments that permit the alteration and augmentation of ourselves. In combination, these two areas of technological and human hardware advancement constitute convergence forces that will continue to offer increased opportunities to *push* more technology into the human sphere.

The first of the two, broad software factors that will drive any increased human–technological convergence arises from the development of artificial intelligence programs. In theory, such technological software may in future develop its own sense of reasoning and even 'identity'. Artificial intelligence software hence has the potential to make the tools with which we work and play appear more and more 'human'.

The second software factor pushing humanity and technology together concerns that basket of *social choices* that continues to be exercised by our *human software* in every decision that leads to the further integration of technological systems into our lives and even bodies. Just because people have the option to share experiences electronically and to augment

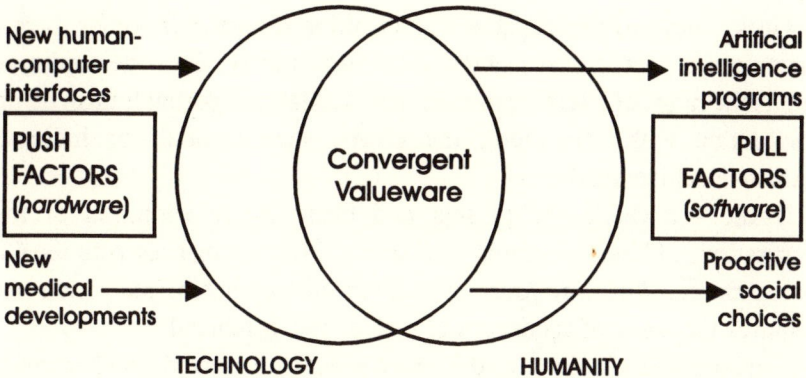

Figure 4.1 **Push & Pull Factors in the Convergence of Humanity & Technology**

their bodies technologically does not necessarily mean that they will. However, the increasing number of private individuals connecting to the Internet does suggest that more and more people may be choosing to foster more and more electronic relationships. There are also already many advocates of the 'transhuman' philosophy which advocates taking a more proactive stance towards further human evolution.

The following sections examine in more detail those technological and social developments that are either pushing or pulling humanity and technology closer together. Partially, the purpose of this analysis is to indicate the value creation implications involved with each hardware or software development. However, in totality, the intention is also to highlight the existence of a radical *Fifth Discontinuity* whose mental bridge we may all soon begin to face.

INTERFACING WITH THE MACHINE

The development of the computer over the past fifty years has undoubtedly been the major factor to fuel the increasing convergence of our machines with our selves. Computers are multi-functional, programmable devices. They are also capable of controlling almost every other form of electronic or mechanical technology. Consequently, the wider and more subtle the range of interface mechanisms available to permit human beings to interact with computers, the greater the potential for increased human–machine synthesis.

Table 4.1 charts the present and likely future evolution of four generations of human–computer interface. This commences with the CLI or *command line interface*—a method of computer interaction that requires the entry of sequences of coded, text commands.

Keyboard-based, command-line interfaces dominated most forms of computing until the mid- to late 1980s. However, since that time, high-power, low-cost computer processing capabilities have been applied to permit CLIs to be overtaken by GUIs (pronounced 'gooies'), or *graphical user interfaces*.

Known to most of the world in the form of *Microsoft Windows*, graphical user interfaces allow people to work with computers largely by interacting with two dimensional pictures or *icons*. These are commonly supposed to represent familiar real-world, desktop objects, and are manipulated by pointing-and-clicking with a variety of interface hardware devices ranging from computer mice to trackballs, touchpads, tablets, and touch sensitive screens.[78]

Graphical user interfaces have undoubtedly made computers far more user friendly. They have hence permitted advanced technologies to be adopted by a wide number of non-technical individuals. The rise of GUIs has also been accompanied by the development of high-resolution computer monitors capable of displaying large, photo-realistic colour graphics. All modern GUIs also feature support for multimedia inputs and outputs, such as sound and video.

Despite their current ubiquity, GUIs are unlikely ever to provide the most fluid means of human–computer interaction. Augmented with voice and optical recognition systems,[79] they will perhaps remain preferable for document creation and administration, and some basic forms of machine control. However, in many areas of three-dimensional design, product testing, simulation, or communication and leisure technology application, GUIs are likely to prove more and more restrictive.

GENERATION	INTERFACE TYPE	BASIS OF INTERACTION
First	**CLI** (Command Line Interface)	**Symbols** (Text commands)
Second	**GUI** (Graphical User Interface)	**Icons** (2-D pictures controlled by pointing-&-clicking within screen windows)
Third	**VR** (Virtual Reality)	**Objects** (3-D representations of realspace manipulated within cyberspace)
Fourth	**Synthesis**	**Human–Technology Fusion** (Direct linkages established between mindspace and cyberspace)

Table 4.1 Generations of Human-Computer Interface
[Developed from *Cyber Business: Mindsets for a Wired Age*, p.42]

STEPPING INTO CYBERSPACE

To compensate for the increasing inadequacy of GUIs like *Windows*, an emerging third generation of human–computer interface allows for the direct, interactive manipulation of *objects* represented in three dimensions in cyberspace.[80] Such VR or *virtual reality* interfaces immerse people within graphics worlds, rather than requiring them to gaze from 'realspace' into cyberspace through a two-dimensional window. VR interfaces hence enable people to interact 'naturally' with data, advanced technologies, and each other.

VR interface hardware usually provides all-round, stereoscopic vision via some sort of head-mounted display (HMD). A dataglove[81] or other three-dimensional gesture device is then employed to enable the *direct* manipulation of the computer-generated graphical environment. Force-feedback systems that enable VR objects to feel solid are also in development. Not much further down the line, full body suits and other forms of advanced, wearable computing hardware will reach commercial application.[82] These will enable large portions of a computer-user's body to be represented in cyberspace. They will also permit human beings to experience a wide perceptual range of interactive feedbacks from those objects and other individuals encountered within virtual worlds.

Admittedly, many of today's media-fantasy-visions of VR remain some years from common application. This point noted, commercial VR interfaces are already enabling the inspection and testing of design prototypes from new automobiles to supermarket stores.[83] Cutting-edge medical VR systems have also permitted surgeons to fly through the innards of their patients in order to assist in the planning of complex operations.[84]

Within prototype virtual offices, geographically separated individuals have donned software bodies to 'meet' over network links. Computer generated mannequins have also paraded new clothing collections down VR catwalks.[85] And, for several years, architects have been using VR to showcase virtual buildings to their clients.

All of the above examples employ currently available VR interface hardware to recreate portions of the actual or potential real world for human occupation. However, other pioneering VR test-beds have moved beyond the interface limitation of modelling realspace in cyberspace. For example, financial information systems have been created that allow their users to fly over VR landscapes of stock and bond prices.[86] Similarly, British Telecom has constructed a VR representation of its entire network through which its engineers may roam at a variety of operating levels.[87]

One of the implications of VR interface development is that it is now becoming possible for human beings to work, play, and do bona fide experiments, in 'surrogate silicon models' of the real world.[88] In parallel, as groupware and collaborative computing become more and more common, 'the human–computer interface is giving way to the human–human interface mediated by computers'.[89]

Durable, lightweight, and relatively low cost VR headsets have already been launched into the personal computer marketplace.[90] These create the illusion of a three-dimensional virtual world by positioning two small screens before a user's eyes. For several years we can expect such hardware to remain the most common means available for visually immersing a human being within a VR environment. However, VR researchers are already devising a far more effective display technology that will allow a visual link to be established between a human being and a computer without the need for an intermediate screen display.

So called 'retinal imaging' uses a low-power laser to scan an image directly upon the back of the eye. As explained by Richard Johnston of the Human Interface Technology Lab:

> With a virtual retinal display we're eliminating the intermediate image or the screen—we're actually using a single beam of light to paint an image on the retina of the eye point by point. Because of this we've got a closer link between the generation of the image, and the view you actually see. This is the first step towards creating a seamless environment between the computer and the human.[91]

Retinal imaging has already been shown capable of producing high-resolution, wide field-of-view colour images. Current prototypes—such as those created by the virtual retinal display (VRD) project at the Human Interface Technology Lab—may be fairly bulky units. However, with new laser scanners in development, lightweight hardware the size of conventional eyeglasses is expected to emerge.[92] Indeed, as far back as 1994, Peter Cochrane, Head of Core Technologies at British Telecom's Martlesham research labs, was predicting the development of unobtrusive, VR contact lenses based upon retinal imaging.[93]

HUMAN–TECHNOLOGY FUSION

In one of his final anthologies, veteran technoculture philosopher Timothy Leary predicted that the main function of the human brain in the 21st century would be 'imagineering' and 'electronic-reality fabrication'.[94] Such a vision may worry many. However, there can be no doubting that computer technologies are becoming more and more closely integrated into our lives. Within markets now championing experiences and relationships as much as traditional goods and physically-manipulative services, there is also an implicit demand for the interfaces that link humans with computers to become increasingly fluid, immersive and transparent. In such a context, it therefore difficult to dismiss out of hand Leary's argument that the key value-creation skills for the third millennium will be those that permit us to 'express, communicate, and share the wonders of our brains with others'.[95]

By using emerging virtual reality interface hardware to access virtual 'realspaces' crafted within cyberspace, many people over the next couple of decades will become used to sharing *experiences* (and not just information) via electronic media. However, the stimulation of our natural senses with external computer interface hardware is almost certain to remain a second-best science. Because of this, we may already identify the beginnings of a fourth generation of human–computer interface. This involves a direct *synthesis*, and will be based upon some form of human-technology fusion.

In the first true human–computer interface paradigm shift, the direct connection of human beings with computer systems will give rise to the primary linkage of 'mindspace' with cyberspace. The intermediate interface metaphor of linking mindspace to realspace *to in turn interact with cyberspace* will subsequently become redundant in some forms of human–computer interaction. With no manipulation of, or stimulation from, secondary interface devices, the hardware of human being and computer will simply become one. Across computer networks, possibilities will hence open up for *directly* interfacing not just people with technology, but also one mind with another via a cyberspace link.[96]

The above claim may sound like pure fantasy. Yet, as long ago as 1974, Dr. William H. Dobelle and his research team at Columbia University were successfully implanting electrodes into the brain of a blind patient that enabled him to 'see' crude patterns. Descendant research projects at what is now the Dobelle Institute will one day enable high quality visual inputs to be fed *directly* into our heads by wired or even non-wired interfaces. Sometime in the 21st century, the day will dawn when we may opt to have a brain interface jack fitted to permit a direct mindspace–cyberspace link.[97]

Progress with non-visual human–technology interface development points the way ahead. Pacemakers are common, whilst the deaf can already have a cochlea implant to enable them to hear via a direct neural feed. Systems also exist for the control of artificial limbs from sensor devices interfaced to their wearer's body. Microfabrication technologies that will allow the low-cost manufacture of silicon chips to interface with our bodies at the neural level are also being perfected.[98]

Frightening choices clearly lie ahead as computer interface hardware developments continue to offer opportunities that may push the conceptual spheres of humanity and technology into a common realm. The option before some individuals may be between non-wired unemployment, or interface-augmentation into a value-rich, experience-drenched super being. However, before we consider the social and software agendas behind such choices, we would be wise first to turn to the flipside of the hardware coin. There may be humanity–technology convergence push-factors emerging from developments in the hardware interfaces of computing machines. However, we must not forget that these are occurring in an age of dawning genetic and medical advancements that may, in parallel, alter the hardware of our own flesh and blood.

REENGINEERING OURSELVES

Across history, human civilizations have long sought to control the characteristics of their population. In particular, if with dubious success, many peoples have tried to influence the sex of their babies.[99] Perhaps because no exact science has existed for controlling our biological make-up, ethical dilemmas surrounding proactive 'human reengineering' have largely been muted. Yet, over only the past twenty years, all this has started to change.

Back in 1978, the first child to be artificially conceived via *in vitro* fertilization (IVF) was born. The IVF process involves exposing collected eggs to sperm outside of the human body, and then culturing the resultant embryo in the lab before transfer back to the uterus. At present, such a procedure to help childless couples remains expensive and is frequently unsuccessful. Nevertheless, IVF technologies have opened the door to a whole host of technological procedures for 'interfering' with human development. IVF hence again calls into question the distinction between the 'natural' and the 'artificial', and the growing interdependence between ourselves and our technologies.

Just one of the implications of IVF is that it is slowly becoming possible to screen the embryos so conceived for some of the potential 5000 single gene defects that may afflict a human child. By 1992, medical technology was making it possible to diagnose cystic fibrosis. Since that time, embryo screening for several other genetic disorders has been developed. Preimplantation screening for some chromosomal abnormalities has also become a reality.

Over the coming decade, an increasing range of invasive and non-invasive screening techniques that will 'read' the DNA or 'genome' of our genetic make-up will become available. In particular, the pending completion of the Human Genome Project[100] is likely to lead to many advances in genetic testing. Launched in the United States in 1988, this fifteen-year quest has the goal of mapping the chemical sequences of the three billion nucleotide base pairs of human DNA.[101] As a result, the Human Genome Project will provide the most complete plan ever of our own, biological 'hardware'. Scientists will thenceforth be able to use DNA testing to predict (and one day even cure) many more genetically-related ailments.

Over the next five to ten years, the ability of medical science to cultivate embryos outside of the human body is expected to increase.[102] In parallel, developments in microsurgery and molecular biology are likely to make more common not just screening for genetic defects, but also genetic manipulations than may result in 'transgenic' human beings.

Transgenic technology permits new genes to be introduced into IVF-cultivated embryos. So-termed foreign genes—or 'transgenes'—were first injected into mouse eggs in the early 1980s.[103] Transgenic animals are subsequently reared by implanting fertilized eggs injected with the desired genetic material into a 'foster' animal. Such a technique was famously brought to the public consciousness in 1996 when it was used to clone a sheep named Dolly at the Roslin Institute in Scotland.[104]

The announcement of Dolly led to fears that human beings would one day be cloned by a similar method. However, in practice it is far more likely that transgenics will be used to breed human–animal hybrids. For example, transgenic pigs have already been bred with introduced human DNA, and can provide genetically compatible organs for human cardiac and various other transplant operations.

Genetic manipulation and so-called 'gene therapy' more broadly may be divided into two distinct areas: somatic and gene-line. The former involves the removal, manipulation and return of patient tissue whose alteration will only ever effect that one patient. This is in contrast to gene-line therapy, whose practice alters the make-up of embryos, sperm or eggs. To some people's alarm, gene-line genetic manipulations will therefore be passed on to every subsequent generation.

The 'reprogramming' of somatic human genetic materials is already assisting in battles against cancer, heart disease, blood disorders, diabetes, and even Alzheimer's. However, at present regulatory authorities do not permit gene-line genetic engineering to take place in humans. This is partly because failure rates remain high and results unpredictable. However, the creation of genetically-manipulated transgenic human beings does remain a technological possibility with powerful implications. As Robert Winston, a Professor of Fertility Studies, contends:

> There are essentially two reasons for wishing to introduce new genes into human embryos. The first would be to correct gene defects which occur in that family, so that future generations would not suffer the particular disease carried in the family. Ultimately there is the notion that specific gene defects could be permanently eradicated. A second objective . . . would be to introduce genes giving specific characteristics which are regarded as desirable. [And] this concept leads to the idea of the 'designer baby'.[105]

THE NEW ALCHEMY

Technological opportunities for influencing the sex, beauty, intelligence, aggression and disease-resistance of our offspring (and perhaps even ourselves via some form of somatic genetic manipulation) may place almost limitless technological opportunities in our innocent hands.[106] Two decades ago respected molecular biologist William Day predicted the genetically-manipulated 'evolution' of a new species which he termed Omega Man.[107] Bruce Mazlish similarly reflects on how technology is increasingly being substituted for biology in human evolution.[108] He also contemplates the resultant emergence of 'prosthetic man'[109] as the 'biological becomes mechanized'[110] in our latest attempt to turn ourselves into machines.

The potential artificial advancement or abuse of the hardware of our own bodies clearly raises a great many ethical debates. These will be returned to towards the end of this chapter. However, before we move on from the hardware push-factors of human–technological convergence, we perhaps ought to note how the gene-splicing developments of the dawning 'biogenetic revolution' are not just occurring *in parallel* with the development of new computer interfaces and manufacturing technologies. Far more significantly, as microelectronics becomes ever more micro, and as medical researchers develop the tricks of the trade of genetic engineering, so we are learning to conceptualize and manipulate the hardware of our machines and our bodies on the same, nanometre scale.[111]

A fledgling science of 'nanotechnology'—of precision manufacturing at the near-atomic level—is already beginning to emerge. A few leading scientists are also predicting that 'nanotech' will become a common tool in both technological *and* human development. According to cutting-edge researchers including Eric Drexler and Richard Feynman,[112] the future of nanotechnology holds in store the creation of molecule-scale self-replicating robots. Some such minuscule 'assemblers' will be able to synthesize both new and existing materials out of the raw components of matter. Other 'nanobots' will even be able to be injected into our bloodstream in order to undertake repairs at the cellular level, or to augment the systems of our bodies by interfacing them with highly-micro human–computer interface hardware.

The construction of the first assembler robots on the nanometre scale will be a challenge that will stretch the very limits of our mental and technological capabilities. It is largely for this reason that nanotech assemblers will have to be made capable of replicating themselves. So the theory goes, build one 'nanobot', give it some time and a supply of common atoms, and a whole army of assemblers will be created to subsequently perform other tasks. Once the first assemblers have been built, their identical offspring will therefore soon become available in their millions and billions. In turn, more and more complex assemblers will be built by other assemblers to fulfil a broader and broader range of functions.

Hardly surprisingly, many people remain sceptical of the claims of researchers such as Drexler, and in particular of his 'bottom-up' theories for the creation of micro-miniature atomic construction robots. However, there is already industrial and financial backing for the less-ambitious development of so called 'top-down' nanotechnology, which involves the production of nanostructures from bulk materials.

As just one example, in 1997 a company called Oxford Nanotechnology was set up to develop production techniques in nanolithography.[113] These will permit the creation of the next generation of computer chips on a scale not attainable with traditional lithographic production methods. In the United States, some observers estimate that over 400 companies have broad nanotechnology interests. Predictions as to the future value and importance of the nanotechnology industry are also substantial. For example, a 1996 UK Government Office of Science and Technology report predicted a market worth £80bn in short-term nanotechnology applications by the year 2000. A European Parliament paper of the same year also commented that highly resource- and energy-efficient nanotechnology could one day 'take the place of present-day industrial manufacturing'.[114]

Without doubt nanotechnology is the most extraordinary science-fiction of the present. Yet perhaps only a few decades hence it will literally become the 'new alchemy' of the future. Indeed, even bottom-up atomic assemblers may become a reality sooner than we think. Eric Drexler, for example, predicts that we are now entering a period of 'superexponential progress' that will culminate in a rapid leap to our absolute technological limits.

When and if this occurs (and it is important not to uphold nanotechnology as a future salvation to all human problems), distinctions between the organic and inorganic world will become largely irrelevant. Too easy to dismiss as impossible, 'nanotech' breakthroughs will hence be many orders of magnitude more significant than those which so recently gave us the extraordinary novelty and power of microelectronics.

DIGITAL HEARTS, COLD MINDS?

Some years ago in a television drama, a scientist and a soldier stood before Colossus—the world's first digital computer, and a key weapon for the allies in deciphering the German Enigma codes in WWII.[115] The scientist was enthusing how in future many more computers would be built as 'thinking machines'. To this the soldier solemnly replied, 'Yes, but whose thoughts will they think?'[116]

The holy grail of creating true artificial intelligence (AI) has been eluding computer programmers for over fifty years. Early promise was shown with software capable of beating a human being at chess. However, it was soon realized that programming a machine to play a logical game, and somehow providing it with the 'intelligence' necessary for general reasoning, were two hurdles of very different magnitudes.

To this day, computers remain good at logic, and hence at automating decision tasks that involve the application of a few, simple rules. However, a step beyond black-and-white decision making, even the most advanced AI programs still exhibit very poor levels of comprehension no greater than those of the average toddler. Human beings may have just about managed to program computers with the capacity to 'read' pages of text via optical character recognition (OCR), and to reliably 'translate' spoken language. However, we have barely made any progress down the road of enabling software to actually *understand the meaning* of those words that we can now scan or speak into our computing machines.

In spite of the above, the lure of AI remains powerful on two distinct levels. From a scientific or business perspective, the potential *application* of software programs capable of 'thinking' for themselves could be enormous. As discussed in **chapter 2**, smart, autonomous software agents are already being created to surf cyberspace as electronic shopping assistants, researchers, credit-authorizers, fault-finders, travel agents, time-schedulers, and more.

Software agents will learn how their human masters behave by studying their patterns of on-line behaviour. From such analysis they will establish databases of rules or 'heuristics' relating to human likes and dislikes, priorities and appropriate actions. Nobody as yet is predicting the imminent development of software agents with any degree of real artificial *intelligence*. This said, over the next decade agents may become so *smart* at cross-linking our every action and reaction that whether they are actually capable of 'thinking intelligently' may become irrelevant. As Professor Donald Michie of the Turing Institute notes:

> If a machine becomes very complicated then it becomes pointless to argue whether it has a mind of its own. It so obviously does that you had better get on good terms with it and shut up about the metaphysics.[117]

The above view noted, the philosophical significance of the potential emergence of true AI ought not to be totally ignored. After all, if and when true AI is brought into existence, so the Fourth Discontinuity will finally cease to exist.

Since their very dawn, human beings have constructed tools to take on the tasks of their bodies. In recent years, construction robots have been fabricated capable of manipulating the physical world with no human intervention. We have hence learnt to technologically recreate hardware with a large degree of human functionality. In turn, if true artificial intelligences are ever programmed, we will also have learnt to artificially mimic the software of our own minds.

The creation of 'thinking' AI programs may at present be the weakest force driving any further convergence of humanity and technology. Yet, when and if effective AI does emerge, it may rapidly become the most powerful. Many computer industry players also believe that AI is no pipedream. As Microsoft guru Bill Gates, for one, once noted:

> Certainly computers will be in any meaningful sense as smart as people at some point. It's turned out to be a very hard problem and it'll still take some time. But we will get there—there's no secret source that the old chips won't be able to deal with.[118]

DISEMBODIED CYBORGS

The development and application of virtual reality computer interfaces, human–technology hardware fusion, and true artificial intelligence programs all lie some years or decades into the future. However, around us today an increasing number of individuals are *already* allowing their minds to touch those of others via the medium of cyberspace. Indeed, perhaps ironically given the widespread popularization of technophobia across much of Western society, the proactive *social choice* to make use of more and more technology is almost certainly the greatest current force driving human–technological convergence.

Large numbers of people now lap up every new technological gizmo within a few years of its emergence onto market. Many also dream of, and pressure for, even more advanced technological hardware and software products. In fact, since our earliest ancestors first made tools with stone and bone and wood, many human beings have been almost *obsessed* with technology and its further integration into their lives.

Such an obsession is today powerfully illustrated by the exponential adoption of the Internet. Already this thriving international network has become not just a business tool, but an information and *social* space adopted by millions of private individuals around the planet. Indeed, many people now spend a significant amount of time on-line. So much is this the case that, in the United Kingdom at least, the uptake of the Internet has led to a doubling in the length of the average phone call.

Already the global hardware platform of the Internet has become a conduit for human selves without bodies. By increasingly sharing their mindspace via cyberspace rather than realspace, many individuals are learning to live on-line—to make friends and to socialize as disembodied, textual entities freed of the boundaries of the physical world. And this is just the beginning. Today the interface of the on-line world is largely limited to that of the GUI. Yet this will undoubtedly change.

At the very least, over the next five to ten years we'll use screens to look *into* three-dimensional graphics worlds when doing business or socializing on the Internet. At most, in less than a decade many people will commonly use some form of head-mounted display to immerse themselves in a VR office, shopping mall, club, or other social space. In all such locations people will inhabit a graphical, software body of their own choosing. At present, meetings on-line may be restricted to interactions via the keyboard. But things *really are* destined rapidly to change.

It is perhaps ironic that many of the first cyborgs—the first hybrids of human beings and machines—will exist in cyberspace as VR software bodies occupied by networked minds. In other words, as the global surfers of the Internet are already starting to demonstrate, human and technological *software* will achieve fusion long before human and technological *hardware*. Cyborg monsters, when and if we create them, are more likely to be found roaming the Internet, than stomping across some mad scientist's lab.

Chapter 6 reports the views of a few of those pioneers who've already felt their lives change as a result of developing an on-line presence and even a cyberspace personality. Whether many protesters like it or not, life as a 'digital nomad' is for a significant minority already the life of the present and the future.[119] Some may bemoan the increasing reach and application of networked technologies. However, they are often missing the point that, as we cease to coddle ourselves in the old prejudices of geography or nation, a safer and more caring, sharing world of richer human *relationships* is already being forged.

It should also be noted that most 'Netheads' are choosing to follow a trend, rather than to create one. Every new e-mail address registered may pull more technological interaction into the human sphere. However, those most radical individuals today driving a proactive coming-together of human beings and technological machines are already trailblazing a far more startling crusade.

THE TRANSHUMAN AGENDA

Some people firmly believe that a proactive stance now ought to be taken in furthering human evolution. Specifically, those who advocate the social agenda of 'transhumanity' contend that we ought to use all available technologies to help develop human hardware and software in all possible directions. To quote transhuman advocate Anders Sandberg, 'transhumanism is the philosophy that we can and should develop to higher levels, both physically, mentally and socially using rational methods'. Noting more specifics, Nancie Clark adds that transhumanism is:

> . . . a commitment to overcoming human limits in all their forms including extending lifespan, augmenting intelligence, perpetually increasing knowledge, achieving complete control over our personalities and identities, and gaining the ability to leave the planet.[120]

The above desires are hardly unique to the late 20th and early 21st centuries. However, transhumanists point out that it is only recently that we have begun to develop technologies that may realistically permit a superior body and mind, wider access to the universe, and greatly extended lifespans. In addition to new human–technology interfaces, genetic engineering, and nanotechnology, the research agendas of transhumanists hence also include the development of cryogenics and even 'uploading'. The former permits human beings (or parts thereof) to be frozen in suspended animation,[121] whilst the latter relates to the potential transfer of a human mind into a computer.[122]

There is, however, more to transhumanism than a simple love affair with conceptual and bleeding-edge technologies. Indeed, above all, it should be appreciated that transhumanism is a *social movement*. Reflecting this, and following a lengthy discussion in an Internet forum, a 'consensus platform' of transhumanist principles was drafted by Alexander Chislenko. This was prepared partly in response to the realization that many of the goals of transhumanity require large-scale collective action. Transhumanists therefore need to develop a coherent message to be explained to the masses. To such ends, their list of transhumanist principles notes that advocates should:

☐ Strive to remove the evolved limits of our biological and intellectual inheritance, the physical limits of our environment, and the cultural and historical limits of society that constrain such 'progress'.

☐ Pragmatically use whatever tools prove effective in pursuit of the above, be they technologies or intellectual disciplines.

☐ Spread awareness of the dangers of technophobia, coercion, antihumanism, and other 'destructive ideologies'.

- ❑ Strive to achieve their own individual ambitions—be these in seeking health, intellect, monetary reward, social success, or political accomplishment.
- ❑ Promote all human efforts to grow and adapt to an ever-changing universe
- ❑ Tolerate all schools of thought that do not seek to limit the extent or variety of individual and collective achievement, whilst discouraging all attempts to impose will or ideas through coercion.[123]

Chislenko's list tends to suggest that the growing movement of transhumanism is already a doctrine that many people would be foolish to totally ignore. Today, serious attempts at life extension may be limited to experiments with lifestyle change, dietary supplements, and meditation. However, as we have noted across this chapter, new silicon and genetic technological tools may already be realistically predicted on the 21st century horizon.

There is clearly a danger that society may become further divided if one group of individuals decides proactively to evolve itself into an overclass of 'post-human' beings. However, it may be equally dangerous to try and pass laws to prohibit further man–machine fusion and transhuman development. Such legislation would not only curtail individual liberties and quality of life, but may even endanger humanity's long-term future. Given the will of virtually every member of the human race to survive—if not to excel—any 'anti-transhumanist' laws would probably also just drive the movement underground.

As the 21st century dawns, technology and humanity look to be increasingly natural as well as symbiotic partners. Proactive social choices are driving such a convergence, with transhumanity in particular likely to be a growing movement of the early to mid-21st century. In some ways this may even be viewed as 'natural'. Indeed as Bruce Mazlish reminds us:

> The first and most fundamental thing that must be said of humans is that they are evolutionary beings. Thus, they are changing beings, and characteristics that might help define them at one stage of their evolution might not be accurate at another.[124]

A FIFTH DISCONTINUITY?

The preceding sections have sought to explain the hardware and software forces pushing or pulling more technology into humanity's previously insular domain. We have hence been investigating those technological and social developments that seem likely to dispel the myth that there will always remain fundamental differences between human beings and machines. Or as noted at the start of this chapter, we have been engaged in an investigation of the bridging of Bruce Mazlish's 'Fourth Discontinuity'.

Centuries ago, renaissance astronomer Nicolaus Copernicus toppled human beings from the epicentre of creation. Many years later, Charles Darwin placed us firmly within the evolutionary gene-pool of the animal kingdom. Not long after, Sigmund Freud dared to steal our delusion of always being in rational control. Compounding these belittling revelations, Mazlish's proposition that a *clear spectrum* now exists between ourselves and 'artificial' technologies may be yet another unwelcome pill for humanity to swallow.

Being forced to once again question our own power and significance can hardly be conceptually pleasant. However, we may perhaps find solace in an accompanying realization that a kinder, Fifth Discontinuity may be starting to be bridged alongside the Fourth. New technological developments and social forces may cause us to question the nature of our humanity. Yet, at the same time, they also interlink new tools and opportunities with which to immortalize ourselves within an emerging, ever-living technosphere.

Reflecting this opportunity, a Fifth Discontinuity may be bridged as we begin to perceive a non-binary scale between human life and human death. Throughout history—and even in the face of intensely powerful religious convictions—'alive' and 'dead' have always remained practical extremes. However, as human and technological hardware and software start to mingle, a startling continuum may be emerging between the previously ordinal brightness of life, and the blackness of the loss of individual existence.

By increasingly interfacing ourselves both with machines, as well as to each other *through* machines, we may inadvertently have begun to blur the life–death discontinuity. Granted, if a human being is defined as a particular slab of meat with an isolated internal software then, yes, she or he will certainly die. However, once we break past the Fourth Discontinuity and put humans and machines on a shared scale, then the whole 'death' question becomes one of several murky shades of grey.

To a perhaps alarming degree, a significant minority of people are starting to become disembodied and part-immortalised across the technological hardware platform of global telecoms and computer networks. To this end, some individuals are already defined as much by *who* as by *what* they are. Business and social relationships with other people encountered not as physical entities, but as patterns of text and electron streams, are becoming increasingly common. Flesh is not quite so dominant the human–human hardware interface it once was. And on-line human beings not *totally* dependent upon organic shells to shape and experience the rush of life need no longer fully perish with their birthtime, biological hardware.

As more and more of ourselves becomes encoded into cyberspace,[125] so death may become more a question of *where* and *if* rather than when. In particular, if the transhuman dream of uploading is ever realized, the ending of a physical life in realspace may lead to the dawn of a disembodied existence in the infinitely-expandable world of the global net. Indeed, even before that time, a trend may develop of leaving a 'living' legacy in cyberspace. I for one would prefer to be immortalized in a perpetual web space with which future human beings could interact, rather than through a slowly-weathered tombstone.

As many transhumanists hope, sometime in the third millennium technology may make it possible for human minds not just to be 'uploaded' into cyberspace, but also to be downloaded back into a replacement, synthetic body. Reincarnation may hence become a computer-mediated reality, in addition to cyberspace the final resting place of some human souls. Certainly, on the brink of the 21st century, it is clearer than ever before that the most effective way to immortalize is to digitize. As a consequence, and as Timothy Leary explained so profoundly towards the end of his biological days:

> In the near future, what is now taken for granted as the perishable human creature will be a mere historical curiosity, one point amidst unimaginable,

multidimensional diversity of form. Individuals, or groups of adventurers, will be free to choose to reassume flesh-and-blood form, constructed for the occasion by the appropriate science.[126]

* * *

LEAPING AHEAD

If you jump into a lake, you don't get wet because it's full of water. Rather, your soaking results from your decision to leap. Similarly, when it comes to interfacing ourselves through and with new technology, we possess a proactive choice. There may be terrific hardware push factors now offering the *potential* for the increased convergence of humanity and technology. However, our decision to pull more technology into our lives will remain just that. For the foreseeable future, we will retain the shoreline option of whether or not to take a running jump.

As we experience the rush of a new millennium, there's an ocean of fresh technological opportunity beginning to boil. The Pandora's boxes of the world's best research labs are unlikely to be held closed so long as markets and capitalism continue to be chosen to mediate our lives. Indeed, it will be the mass participation of individuals in economic transactions that will determine the degree to which we decide to bathe in new, technological waters.

Whatever many people may fear, technology alone will never do anything to our humanity. Hardware push may soon offer incredible or horrific cyborg potentials. However, software pull will always remain the dominant convergence force. Technology alone will never do anything to our humanity. Though increasingly, we may do choose to do things to our humanity with technology—to leap into the exhilarating chill of the only-partially-known with a passion that has ceased to worry about getting wet.

The three chapters of **Part I** have investigated a wide range of interlinked technological, organizational and social value engines and convergence forces. All, I believe, are destined to impact heavily on our lives in either the short- or medium-term. To explore just how this may happen, the following three chapters of **Part II** report a wide range of beliefs and expectations concerning the nature of future value and its creation.

More than anything, I hope that those chapters ahead provide some measure of those choices—both collective and individual—whose enactment will shape the world in which we all must live. Conclusions, so far as they can be drawn, I leave to **Part III**. So first, let us walk to the edge of the water, and attempt to see just how far out we may very soon choose to jump.

PART II
VALUE PERSPECTIVES

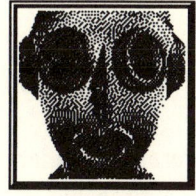

5
Maximising
Corporate Success

OVER LUNCH AT A recent industry workshop, a senior manager complained how his company often suffered from 'in-flight magazine strategy'. Another attendee from a different organization nodded in sympathy as he ate. A third then put down his drink and bemoaned the consequences of his boss having gone off on holiday with one of the latest best-selling business books.

As a writer of business-slanted, future-gazing texts, I perhaps ought to be careful in any apparent dismissal of the latest fix-all management tomes. However, looking over from that side of the fence haplessly charged with finding and preaching the Next Big Thing, I find it fascinating to witness the relentless appetite of both students and practising managers for new rules or philosophies that will 'guarantee' future value creation. The latest sparkling wisdoms packaged as sure-fire business certainties by academics and guru consultants do usually dull very rapidly. However, this does not seem to stop a large number of businesses coming back for more from those of us constantly sought out and then rejected as 'always being wrong'.

Recently I started supervising yet another MBA dissertation. The topic was in an interesting but relatively new field in which little had been written. To my new student's dismay, I was therefore unable to provide a substantial list of initial references. As he subsequently only half-joked in complaint, 'I came here like a hungry bird with its mouth open ready for you to throw in the worms. And you're telling me that the worms just aren't available!'

In parallel with the rise of management as a profession, the past one hundred years has been witness to the practice and publication of a myriad of management theories for those charged with maximising corporate success. **Table 5.1** summarises just eight of the most distinct and influential. In doing so, it also provides a route map for this first of three chapters that will explore a wide spectrum of value-creation perspectives.

Whilst **table 5.1** does provide a common frame of reference, readers should note that none of those populist management approaches or 'waves' it lists are being presented here as completely 'right' or totally 'wrong'. Some theories remain suited to past market conditions, whilst others provide little more than a new lexicon to describe old problems. Some even, perhaps, simply report good business common sense.

However, in amalgamation, those management approaches discussed within this chapter do at the very least highlight the most common worms to have yet been unearthed to feed the open mouths of hungry, value-starved organizations. They hence reflect the evolution of what has been *thought* to be the best business practice for optimal value creation past and present. Granted, this does not necessarily mean that the management ideas listed within **Table 5.1** reflect actual past or present *practice*. However, if only given the long-term sticking power of many of the approaches, it would be foolish to dismiss all eight completely out of hand.

THE ONE BEST WAY

Back in the late 19th and early 20th centuries, many organizations were growing in both size and complexity. Partially as a result, problems of poor coordination became a common management concern. In response, several 'systematic' management methods were championed. These highlighted the need for carefully defined duties, responsibilities and work structures, standardized techniques, and clear cost accounting.[127]

In many respects, as a 'Fordist' logic for the mass production of standardized goods began to overtake the previous industrial paradigm of craft-based manufacturing, so the rise of highly regimented, rule-based management techniques was virtually inevitable. Indeed, the only way for pioneering entrepreneurs like Henry Ford to spawn new large-scale manufacturing industries was with rigid assembly-line techniques that effectively turned human beings into the cogs of great technological machines.[128]

APPROACH	FOCUS	BASIS OF VALUE CREATION
Scientific and administrative management	Rules and structures	Rules, routines and structures for optimal output and work efficiency.
Human and neo-human relations	Individuals	Considering the social and psychological needs of workers to improve their productivity.
Systems and contingency theories	Environmental interaction	Interpreting the organization as a socio-technical system within a wider environment.
Excellence linked to corporate culture	Organizational culture	Fostering a strong holistic atmosphere of shared values to achieve excellence.
Total quality management (TQM)	Quality	Values and quantitative systems to enable the very best products and services to be produced.
Value chains and business process reengineering (BPR)	Customers and processes	Focusing on customer needs at each stage of every organizational process.
Value based management (VBM)	Stakeholders	Balancing the needs and rewards of shareholders, workers and customers.
Knowledge management	Organizational learning	Profiting from experience and encoding core competencies into a living organizational fabric.

Table 5.1 Management Approaches for Corporate Success

One of the most influential management pioneers of rising Western industrialization was Frederick Winslow Taylor. His particular contribution was to develop various principles of 'Scientific Management'. These were devised to increase productive efficiency and to reduce waste, and most notably included:

- ☐ The replacement of rule-of-thumb guidelines with a scientific approach for finding the 'one best way' to undertake any individual work task.
- ☐ The scientific selection, training and development of workers in order to allow them to labour optimally in the 'one best way'.
- ☐ The separation of 'brain power and muscle power' so that 'managers would manage and workers would work'.
- ☐ Management/worker cooperation to ensure a match between task planning and execution.

On the face of it Taylor's philosophies may have seemed reasonable enough. Indeed in some famous cases they reaped major results. For example, Taylor managed to scientifically select and train some of the pig-iron handlers at the Bethlehem Steel Corporation so as to increase their average daily load of metal moved from 12½ to 47 tons.[129]

Such efficiency 'triumphs', however, were not without their price. For a start, Taylor's time and motion study methods, together with his stated desire to 'stop the loafing' of those manual workers with such a 'limited intellectual capacity', led to accusations that he was inhuman and intent on workforce exploitation. Indeed, Taylor's engineering-inspired management methods were investigated by a House of Representatives Special Committee in 1911 following worker unrest after an attempt to introduce Scientific Management into a government arsenal.

Other early proponents of scientific- or administrative-led management included Henri Fayol (who developed 14 principles for effective management),[130] and Frank and Lillian Gilbreth (who refined Taylor's methods for the factory floor). Along with Taylor, all such advocates of what is sometimes referred to as the 'task school' undoubtedly made significant advances in improving the efficiency of people as machines. However, their approach also left the greater human potential of many workers unexplored.

HUMANISING THE LABOUR MACHINE

By the early 1930s, the assumption that most individuals were best managed as rational economic machines was starting to be challenged. Most famously, a set of detailed investigations conducted from 1924–1932 at the Hawthorne plant of the Western Electric Company turned up some controversial results.

In the first of four experiments, Elton Mayo and his Harvard team altered lighting levels in part of the Hawthorne factory in an attempt to establish a scientific relationship between expenditure on lighting and productivity. However, no systematic link between these two variables was found to exist. Indeed, on occasions productivity actually went up when illumination levels fell. From this, the researchers later concluded that workers in the factory were behaving differently as a result of the interest being shown in them by the Harvard team. The implication that psychological and social factors could have a significant influence upon human productivity subsequently became known as the 'Hawthorne Effect'.

Mayo *et al* next observed six women assembling electrical relays over thirteen test periods and under various physical working conditions. A close-knit, supportive social atmosphere was noted between the women throughout, and with all but one change in working conditions their productivity increased. However, once again no clear causal connection could be established between changes in the physical work environment and employee output.

In the third phase of investigation, over 20,000 interviews were conducted in order to gauge employee attitudes to their supervisors and the workplace. Once again these appeared to motivate staff. By now suspecting that the human interest of the researchers was the main factor influencing productivity increases, Mayo and his colleagues finally observed fourteen men assembling terminal blocks under normal factory conditions. Unlike in previous experiments, care was taken to minimise social contact between the researchers and the workers. Over time, strong social pressures were once again observed within the work group. However, in this instance these actually served to restrict group output to a level the men jointly considered to be 'reasonable'.[131]

From the 1940s to the 1960s, further theories of 'Neo-Human Relations' were developed. These attempted to make a more concrete science out of the Hawthorne message that human motivation and group output could be functions of positive and negative psychological and social processes.

Abraham Maslow, for example, linked motivation to a 'hierarchy of needs'. This suggested (with some empirical validity) that people's motivations changed as certain levels of discrete need were fulfilled. For example, Maslow argued that once someone's 'lower level' sustenance and shelter needs had been met, their motivation could only be improved by appealing to higher social desires such as group belongingness or self-esteem.[132]

The theories of Frederick Herzberg and Douglas McGregor similarly suggested that managers ought to consider the motivation of workers on a particular needs basis. Herzberg hypothesized that some aspects of the work environment only constituted *hygiene factors* whose *absence* would demotivate people, but whose improvement would have no impact on output. Hygiene factors therefore needed to be isolated from *motivators* (such as praise) that could be manipulated to improve productivity.

Taking a slightly different stance, McGregor argued that people at work fell into two distinct attitude sets which each required a different management style. Under McGregor's 'Theory X', people viewed work as drudgery, and hence managers had the task of persuading and coercing them to improve their output. Alternatively, under 'Theory Y', McGregor suggested that other individuals viewed work as a potentially-fulfilling opportunity. This second set of workers would therefore be more greatly motivated by managers who facilitated a joint, teamworking mentality of 'showing them the way.'

BROADENING PERSPECTIVES

In line with scientific and administrative management techniques, human relations approaches exhibited a narrow, internalistic frame of reference. For Taylor and his followers, the tools for business success had been those of isolating a formal 'one best way', and then providing the incentive of a higher financial reward to motivate employees. Almost at a polar extreme, and in their attempts to improve further individual and group productivity, advocates of human relations had then focused almost exclusively on the informal, social and psychological needs of workers.

Given the above, it is perhaps hardly surprising that later management approaches were to seek more open conceptual horizons. In particular, from the 1950s onwards, various attempts were made to understand organizations as combined technological *and* social entities within a wider business environment. The holistic perspective of such *systems theories* was to view organizations as amalgamations of several interacting and interdependent sub-systems. This approach was clearly conceptually, rather than practically, founded. Nevertheless it provided a few notable contributions to the art and science of management.

Firstly, systems theories highlighted the need to consider the impact of technological change upon social structures and human motivation. In one famous example, the potential importance of considering organizations as 'socio-technical systems' was demonstrated following the introduction of new machinery in British coal mines.

Researchers Trist and Bamforth were called in when the new machines were deemed not to be yielding their expected productivity gains. They soon noted how the 'longwall' work patterns introduced alongside the new automated coal-cutters and conveyors no longer required the small, tight-knit work teams characteristic of the previous 'short-face' method of coal production. Operation of the new, large-scale machinery across a thirty-yard coalface had therefore significantly reduced social interaction underground. Labour productivity had subsequently plummeted.

As a solution, Trist and Bamforth devised a 'composite longwall' way of working which permitted small, cohesive groups of miners to operate the new technology. This eased the social problems caused by the introduction of the modern coal-cutters and conveyors, and productivity soon rose.[133]

Aside from signalling the importance of managing the *interaction* of an organization's technological and human resources, systems theories also focused on the need to view organizations as *open systems*. This involved conceptualizing organizations as part of a wider environment from which they received raw material inputs, and to which they sought to return valued outputs.

Into the 1960s, and in the wake of the rising open systems organizational perspective, various *contingency theories* were put forward. These suggested that individual organizations needed to adapt their strategies, structures and internal systems to best suit the characteristics of the environments they faced.

For example, in 1961 Tom Burns and G.M. Stalker made a study of the organizational structures and environments of a range of Scottish manufacturing firms. From this they concluded that the structures exhibited by organizations could be placed on a spectrum ranging from those that were rigidly bureaucratic or 'mechanistic', to those that were far more flexible or 'organic'. They then suggested that mechanistic organizational structures were best suited to firms with relatively stable environments, for example as observed in textile manufacturing. In contrast, organic structures were deemed essential in more volatile industrial climates, such as consumer electronics.[134]

Further studies on both sides of the Atlantic argued that not only organizational structures, but in addition broad strategies and internal organizational processes, ought to be contingent upon the characteristics of an organization's environment.[135] Contingency approaches were therefore potentially powerful in highlighting the strengths and weaknesses of some organizations in relation to their markets. However, in line with earlier socio-technical systems theories, the contingency approach tended to remain descriptive rather than practical. It also never really provided any concrete management tools.

EMULATING EXCELLENCE

In 1982 one of the most successful 'how to' management books ever written was published by Tom Peters and Robert Waterman. Entitled *In Search of Excellence*,[136] it reported 'lessons' following an in-depth study of forty-three of America's best-run companies.

Back in 1977, a general concern with the problems of modern management had led to the creation of a task force at McKinsey & Company in order to 'go back to the drawing board on organizational effectiveness'.[137] Peters and Waterman were the leaders of this project, and set about isolating a number of large organizations whose growth and financial performance signalled them as 'excellent'. Their intention was to study these organizations by a range of means—including interviews with their top executives—in order to isolate any significant management rules or approaches that the likes of IBM, 3M, Procter & Gamble, Delta Airlines, Hewlett-Packard, Johnson & Johnson, Boeing, McDonalds *et al* had in common. In particular, many analysts outside of the task force hoped that this work would clarify a clear, contingent relationship between an organization's strategy and structure.[138]

Peters and Waterman, however, went far further. Backed up by four years of studies, they put forward their now famous '7-S' framework to encapsulate those seven variables that 'any intelligent approach to organizing' had to encompass and to treat as interdependent. This framework helped in forcing explicit thought not only about the hardware of any organization—its *Strategy* and *Structure*—but also about its software of *Style*, *Systems*, *Staff*, *Skills* and *Shared values*.[139]

The key conclusion of the McKinsey study linked organizational success to the existence of a strong set of shared values or 'corporate culture'. Indeed, it was concluded that there were no hard-and-fast scientific or financial 'hardware' rules that all 'excellent' organizations exhibited. Rather, a powerful holistic atmosphere pervaded at all levels and gave rise to 'excellence'. The management of a strong corporate culture was therefore championed by Peters and Waterman as the key to present and future organizational success. As in the Hawthorne Studies over half a century earlier, researchers in search of 'hard', scientific rules had concluded that 'soft' processes had a more significant role to play in optimizing value creation.

Peters and Waterman provided a list of eight attributes which they claimed were characteristic of a corporate culture that did and would produce excellence. These encapsulated their observations that excellent firms 'were brilliant on the basics', 'didn't substitute tools for thinking', 'worked hard to keep things simple in a complex world', and allowed their champions 'long tethers'.[140] For Peters and Waterman, the route to optimal value creation thereby lay in imitating a corporate culture which espoused:

□ A bias for action
□ Sticking close to the customer
□ Fostering autonomy and entrepreneurship
□ Achieving productivity through people

□ Being hands-on and value driven
□ Sticking to the knitting (*focusing on the core business alone*)
□ Exhibiting a simple form with a lean staff, and
□ Simultaneously having loose–tight properties (*pushing autonomy to the shop floor, whilst being fanatically centralist about core values*)

Even though the Peters and Waterman message was powerful, learning how to profit from it proved more problematic. As other researchers have since agreed, corporate cultures are 'manifested' in the form of organizational symbols, heroes, rituals, values and practices.[141] The 'contents' of strong corporate cultures may hence include organization-specific artifacts, language, behaviour patterns, heroes, symbols, beliefs, values, history and ethical codes.[142] Almost inevitably, therefore, corporate cultures remain both difficult to measure and to manage—even though the manipulation of such a 'soft' thing has the potential to deliver such 'hard' results.[143]

Another stumbling block for the excellence–culture paradigm arises from the fact that many of the companies in the Peters and Waterman study have since lost their excellence. IBM, for example, was turning in multi-billion dollar losses—the largest in corporate history—by 1992. Many managers and academics have therefore suggested that the analysis reported in *In Search of Excellence* was too static. Indeed, even the book's authors now contend that no single set of prescriptions can always provide an adequate or even appropriate management toolkit. This said, Tom Peters does still signal the need for an innovative, close-to-the-customer action bias in any battle to achieve corporate success.

QUALITY FIRST

Whilst Peters and Waterman and their followers were focusing upon the achievement of excellence via the manipulation of the 'intangible social glue'[144] of corporate culture, other management gurus were arguing that the key to organizational success lay in focusing upon the *quality* of the goods and services a company produced. Indeed, J.M. Juran, W. Edwards Deming, and other advocates of so-termed 'total quality management' (TQM), had been arguing for years that only clear, quantitative methods could stop poor quality happening. They also suggested that TQM methods could be equally applicable to both manufacturing and service personnel.

Whilst TQM only started to become popular in Western economies in the 1980s, its practice had been widespread in Japan since the 1950s. To this day, many Japanese firms still compete for the annual 'Deming Prize' for quality progress. Indeed, almost ironically, it was the world-beating industrial might of high-quality Japanese competition—nurtured using TQM methods developed for the USA—that, in the 1980s, finally drove many American and European organizations to pay more attention to TQM and just-in-time (JIT)[145] manufacturing.

A key premise of TQM is that it is better to produce goods and services of a high quality in the first place, rather than to implement mass inspection systems to isolate quality defects further down the line. As Deming notes, under TQM managers should seek to create a constancy of purpose leading towards an ongoing improvement of products and services. However, at the same time, they should also back this human resolve with systems that produce statistical evidence to show how quality is monitored and built into their production process.[146]

As in contingency and excellence–culture management approaches, TQM clearly strives to build a consideration of customer (environmental) needs into its value creation philosophy. Indeed, a common slogan for TQM is that of the 'customer being king'.

Advocates of TQM are also at pains to point out that we now live in a new economic age in which delays, mistakes, defective materials, and defective workmanship, are no longer commonly accepted. For some individuals and organizations, price has become a second-order value-marker, with quality instead used as the comparative benchmark in an increasing number of purchase decisions. Cheap products and services are all very well. However, if they frequently prove defective, then the companies that produce them will go out of business.

REENGINEERING ACROSS THE VALUE CHAIN

In an attempt to take the customer- and quality-focused regime of TQM one stage further, in recent years many organizations have sought to reinvent their most basic of operations. Such *business process reengineering* (BPR) draws to some extent from the work of Michael Porter, as discussed in **chapter 1**. This is because it encourages firms to gain a competitive advantage by ensuring that value is added at *every* stage of their productive process or 'value chain'.

It should be noted from the outset that BPR is far more fundamental than TQM. This is because BPR involves starting over to do things better, rather than 'merely' seeking incremental quality improvements in *existing* processes and value chains. Indeed, as process redesign gurus Michael Hammer and James Champy explain, BPR may be defined as a:

> . . . fundamental rethinking and radical redesign of business processes to achieve dramatic improvements in critical contemporary measures of performance, such as cost, quality, service, and speed.[147]

BPR begins with no assumptions and is 'fundamental' and 'radical' in that it demands more than superficial changes in order that customers will become the primary focus of every stage of an organization's value chain:

> Customers, after all, don't care about how organizations achieve their outputs, who is involved, or within what structure they labour. All customers care about are the final products and services that an organiz-ation delivers. Yet for many organizations, rethinking their business processes—their value chains—so that they are totally customer driven can be [most] painful.[148]

In championing new, more *effective* and well as more *efficient* business processes, BPR initiatives frequently involve the use of new information technologies (IT). As already explored in **chapter 2**, today networked computer hardware and new middleware software can play a crucial role in redefining the means by which business is undertaken. These new IT developments may also be employed to alter the interface used to mediate the relationship between customer and organization.

As Hammer and Champy argue, IT has become an 'essential enabler' of BPR.[149] Thomas Davenport and James Short go even further with the claim that IT serves not just as a tool but as a *recursive force* within BPR initiatives. In explanation, they suggest that BPR not only leads to the pulling of various revolutionary 'IT levers'. In addition, they contend that once new IT systems are in place, they will then catalyze even more radical possibilities for supporting customer-focused process change.[150]

SHAREHOLDER & STAKEHOLDER VALUE

Ask many a manager today what their ultimate objective is, and they will reply with the stock phrase of 'maximising shareholder value'. This reflects the sad fact that we live in an age in which organizations are seen more and more as units in which to invest, and less and less as productive systems that exist to combine human beings and technologies in order to create valued products and services.

Shareholders—be they pension funds or single individuals—legally and financially *own* organizations. Organizations that under-perform in their shareholders' eyes experience a reduced share price, lower investment, and may risk being taken over or even going out of business. In contrast, organizations that perform well on the stock markets earn their share-holders high dividends, increase the monetary value of each shareholder's investment, and are likely to continue to survive. Many managers therefore have an incredible incentive to run their organizations for the maximum benefit of their shareholders, even if this may on occasions be to the detriment of their employees, other organizations, long-term capital investment, and society at large.

Fairly recently, however, the profit-first doctrine of running organiza-tions purely for short-term shareholder value maximization has started to be challenged. Partially in its place, a philosophy entitled *value based management* (VBM) has been developed in an attempt to breed reward systems and corporate cultures that address the wider needs of an organization's 'stakeholders'.

Stakeholders are those people who do not necessarily own part of an organization, yet who are influenced by its management, strategy and operations. Stakeholder groups therefore include an organization's workforce, customers and suppliers—and indeed everyone who lives in the wider environment that the organization may change or pollute—*in addition* to its owners. Organizations that seek a maximization of stakeholder value will hence try to meet their shareholders' needs, though not always exclusively in the short-term.

Many companies now cite value based management as their core philosophy. Hyundai, for example, claims to exist for 'mankind, society and the future' by being:

> . . . wholly dedicated to its duties and obligations as a corporation by providing better products and services for customers, and a better workplace for its employees so as to develop the society where we all live.[151]

Just as many companies are now preaching value based management, so many organizations also claim to have invented it. The Deloitte & Touche Consulting Group, for example, states that the foundations of VBM were laid over a decade ago by its strategy consulting group, Braxton Associates. Deloitte Touche goes on to explain how companies today have to balance the 'two imperatives' of satisfying customer needs better than competitors, whilst at the same time delivering superior returns to their shareholders. 'Focusing on either goal to the detriment of the other leads eventually to failure . . . [and] . . . Value Based Management enables organizations to meet this dual challenge effectively'.[152]

The Deloitte Touche Consulting Group takes clients on a 'value journey' in order to introduce a 'massive cultural shift' into their organization so that 'no one will waste resources on activities that simply do not matter'.[153] Critical in this journey is an assessment of a company's value maximizing goals and objectives, strategy planning and resource allocation, operational decision making, quantitative performance measurement, and employee compensation and reward schemes.

The significance of effective worker rewards and motivation in VBM is also highlighted by the Center for Economic and Social Justice (CESJ) in Washington. This too claims to have developed VBM, which it describes as 'a business philosophy and management system for competing effectively in today's global marketplace, centered around the inherent value and dignity of every person'.[154]

According to CESJ, 'businesses are recognising that, for their own survival, they must find new, more flexible ways of rewarding and motivating their workers, while controlling costs and delivering ever-higher levels of value to their customers'.[155] They may do this by crystallizing a leadership philosophy around a set of universal moral principles, as well as providing incentives in the form of share ownership plans that will give workers an increased financial stake in their organization. VBM hence involves a 'free enterprise version of economic justice', and starts with a 'recognition of the value of each person—each customer, each supplier and each worker'. By empowering workers and management with a foundation of shared ethical values, and a reward structure based on the value individuals contribute, so customer and shareholder value will also be maximized. As CESJ assert:

> Experience within a growing number of companies indicates that the more that people's self interests are unified within a management system reflecting the principles of Value-Based Management, the greater customer and employee satisfaction will be. From this can flow increased cost savings, increased sales, and increased profits.[156]

Clearly VBM philosophies and techniques draw from a combination of the best of scientific management, human relations approaches, systems theories, and the excellence–culture paradigm. A little customer-focused TQM and BPR is also thrown in to ensure that shareholder value will be maximized in the long-term. However, given the way in which the best of the 'old' management approaches usually permeate the new, the development of such an umbrella approach is hardly surprising.

THE LEARNING ORGANIZATION

Alongside a philosophy of VBM, many organizations now cite effective *knowledge management* as critical for their long-term success. Corporate expenditure in this area is also growing fast. Dataquest has predicted that in 1999 businesses will spend $4.5bn on improving their utilization of knowledge resources, whilst the Cranfield School of Management forecasts that 6.2 per cent of all business revenues will be devoted to knowledge management around the year 2000.[157] So just what is this increasing amount of money being spent on? And is knowledge management an effective future panacea for optimal value creation?

Like many other 'Next Big Things', knowledge management tends to mean different things to different organizations. This said, most knowledge management initiatives seek to combine adaptive, new technology systems (such as intranets) with the cultivation of a corporate culture that recognises the value of shared knowledge as a competitive asset.

The stated goal of the majority of knowledge management initiatives is to leverage past and present human experiences *through* new technology systems in order to encode learned competencies into the fabric of an organization. Indeed it should be firmly noted that knowledge management is not just about the effective application of new IT systems. Nor—despite woolly definitions in many quarters—is it solely a new cultural philosophy. Knowledge management at its best combines both human and technological hardware and software in the hope of synergistic value paybacks.

Critical to any effective understanding of knowledge management is an appreciation of the difference between 'knowledge' and 'information'. Both terms are still used interchangeably in many quarters. However, their perhaps subtle distinction is highly significant in value terms. In a strict sense, information is the product of filtering and then processing raw *data* into a potentially useful form. Another processing stage on, knowledge may then stem from an analysis of selected *information* within an expert frame of reference so that it becomes attributed with actual *meaning*.[158]

Today, many managers find themselves overloaded with information. Yet amid the constant electronic and paper influx of figures, facts and opinions, knowledge often remains a scarce resource. Addressing this problem, knowledge management is intended to extract more useful knowledge from the mass of information that technological and human interactions continue to produce. In the process, by encouraging a philosophical sea change from information push to information pull (as discussed in **chapter 2**), good knowledge management ought to actually ease the burden of information overload.

Already many organizations have combined new information technology systems with a more open, knowledge-sharing culture to great effect. To cite but a few examples, by improving the management of its product information communications processes, Silicon Graphics managed to reduce its sales training costs from $3 million to $200,000. The Dow

Chemical Company's Intellectual Asset Management programme has also found savings of some $40 million from a full analysis of the patents in its portfolio. Improved knowledge management has allowed Skandia Insurance to cut the start-up time for a new office in Mexico from seven years to six months. And Texas Instruments has avoided the cost of an entire $500 million silicon water fabrication plant by leveraging internal knowledge and best practices.[159]

Often associated with knowledge management is the concept of creating a 'learning organization'. This idea was first developed in the 1980s by Richard Pascal. It was then popularized in 1990 by Peter Senge in his book *The Fifth Discipline*.[160] As David Birchall and Lawrence Lyons have also since explored to great effect, many organizations are now looking beyond individualized training and development in order that their company *as an entity* may develop its own encoded knowledge and learning capabilities.[161] This may enable an organization to cope most effectively with environmental change, as well as to optimize the value of its own core competencies and intellectual capital.

'Learning to learn' also potentially leaves an organization less vulnerable to changes in its human resource base. Given the increasingly flexibility of many labour markets (as explored in **chapter 3**) this may prove highly significant in value terms. As Paul Pederson of Price Waterhouse comments 'every year a significant amount of experience walks out of the elevator. There is thus a need to capture this experience so that new hires can get up to speed very quickly'.[162]

* * *

MANAGEMENT BY ENVELOPES?

To close his book *Technotrends*, Daniel Burrus tells the tale of an incoming manager of a 'large but shaky' corporation. His predecessor has left him with a bottle of scotch, ten envelopes, and the instruction to open an envelope if and when trouble arises.

After only a few months a disaster strikes and the first envelope has to be opened. Inside is a message proclaiming 'Re-engineer everything!' Taking this advice, the manager instigates a major BPR programme and, sure enough, things improve. But not for long—and so the second envelope is soon ripped open to reveal the message 'Establish a learning organization!'

This piece of advice also initially yields results and then fails. The same pattern is repeated following the third envelope's message of 'Benchmark!', the fourth's advice of 'Empower your employees', through to the seventh's of 'Total Quality Management', and the ninth's of 'Quality Circles'. Indeed, nine times in a row an apparently excellent fix-all solution first delivers powerful results, and then for some reason starts to falter.

Despairing as his market share evaporates, the unfortunate manager drinks the scotch and plucks up the courage to open envelope ten. For once the piece of advice inside does not promote a new value creation ideology. Instead, the final message from his predecessor simply reads 'Prepare ten more envelopes'.[163]

As the end of this chapter approaches, it would be nice to conclude that over the decades a clear paradigm shift has taken place in management theory. This *could* be theorized to incorporate a shift from quantitative rules to qualitative values; a change in emphasis from efficiency and output quantity to effectiveness and output quality; and perhaps even a management attitude evolution away from treating workers as economic machines and towards nurturing them as holistic stakeholders.

Such a neat theory may at first seem both plausible and attractive. However, anybody attempting to champion this academically contrived view would be somewhat naïve. Not least this is because the above implies that managers as *people* have fundamentally changed across the decades. And I, for one, would never claim this to be the case.

These points noted, what we may perhaps recognise is how the *emphasis* of management techniques and value perspectives has broadened over the past century in a pattern that has in some ways come full circle. Indeed, one may reasonably argue that knowledge management is simply a revisited form of scientific management to once again champion the encoding of the competencies of the 'one best way'. In fundamental terms, the only real difference between scientific and knowledge management is that the latter is at pains to highlight the synergistic importance of human as well as technological valueware. Though given that Elton Mayo highlighted this fact way back in the 1930s, the humanization of encoded knowledge endemic to a learning organization should hardly strike us as a revelation.

Across history, the best of an 'outgoing' management approach has almost always been taken on board to become part of the next wave. Such a trend is also likely to be ongoing. Management as a science will therefore continue to get more analytically effective, whilst management as an art will increasingly learn to understand the human animal as both a worker and a customer. Value-based management and knowledge management may be the Next Big Things of the present and near future. However, like previous waves of thought and practice, they remain small conceptual stepping stones on a pathway of continual managerial and organizational understanding.

As we may be reminded by Daniel Burrus' excellent business fable, no single management approach—however tried-and-tested or Next Big Thing—will ever prove a panacea for sustained, long-term value creation. Management by envelopes will therefore remain as good a philosophy as any.

More significantly, and as also highlighted by Burrus' tale, organizations usually stay ahead of their competition by doing *something*, rather than accepting the decaying certainly of the status quo. As soon as anything 'new' comes along, it therefore needs to be grabbed or followed fairly rapidly before its ubiquitous application turns first-mover, new-paradigm competitive advantage into a must-have organizational essential. In business, as in biology, the only way to survive and thrive is to embrace the pain of continual evolution.

6
Voices from Cyberspace

You were so far from me
You were ash dead
And your warmth was kept within;
So like a fire that is shielded,
I felt the cold
And sensed the heat.[164]

WHETHER THEY USE IT or not, most people have a definite opinion of the Internet. Some 'netheads' freely admit to being hooked. However, the majority of users seem either to view the Internet as a great *time saver*, or else as a *time-sucking* technology that demands far too much attention in their already busy lives. In particular, many, many managers today complain about the demands of having to deal with a deluge of e-mail messages. As one manager in British Telecom—the company that promotes the phrase 'work smarter, not harder'—recently told me, once he's cleared his e-mail it can often be mid-afternoon.

Non-users of the Internet similarly tend to be somewhat polar in their opinions. Many still simply dismiss the Internet as a fad, or else as another of those 'sad', insular, anti-social hobbies enjoyed by people in anoraks who were once quite content to be train spotters. However, recent market research also informs us that another, sizable proportion of non-users clearly recognise the value of the Internet, if perhaps in a mysterious way that they don't yet quite understand.[165]

Journalists, technologists, researchers and academics tend to be even more clearly divided when it comes to the pros or cons of cyberspace.

Many, including Bill Gates, Nicholas Negroponte, Douglas Rushkoff, Don Tapscott, Howard Rheingold, and Timothy Leary, harbour the rosy, positive belief that new digital frontiers ought to be embraced as they have an enormous potential to enrich human experiences and lives. However, other influential writers, notably including Mark Slouka and Allucquère Rosanne Stone, appear to see little of value—and a great deal that is potentially dangerous—in our continued embracement of the cyber domain.

This second chapter of value perspectives attempts a balanced evaluation of five common debates often associated with the dawning of the 'Wired Age'. Specifically, the following pages try to ascertain whether cyberspace *really is* a new frontier; whether there is a danger of addiction to a disembodied, networked existence; the implications of the point-and-click immediacy of electronic exchange; how a trade in networked relationships may offer an alternative to unsustainable consumption; and finally whether the Internet is already taking on a life of its own.

The diverse range of opinions reported within the following sections reflect the musings of a broad spectrum of fellow futurists, technologists, business gurus, and academics. Some are views expressed off-line in leading books. However, others have been gathered first-hand from e-mailed questions and on-line discussions held within pioneering virtual communities.

More than anything, what the following range of opinions most powerfully demonstrate is how we may learn a great deal about future value, and its creation, from those communities of wired individuals who already inhabit the Net. Something extraordinary *is* now happening on-line. Indeed, the renaissance dawn of our embryonic technoculture may provide the best guide yet to the human value systems of the 21st century.

THIRDSPACE?

Artificial reality guru Myron Krueger once explained that 'for centuries, the goal of human effort was to tame Nature's terrible power'. He then went on to ponder how 'our success has been so complete that a new world created by human ingenuity has emerged'.[166]

The world that Krueger was referring to is now most commonly referred to as 'cyberspace'. This conceptual, electronic universe is argued by its advocates to exist within and between humanity's interconnected computing and telecommunications machines. The 'virtual world' of cyberspace has also been labelled as the next great human frontier.

The term cyberspace was first coined by science fiction author William Gibson in his novel *Neuromancer*. Within, he described the on-line world as that 'consensual hallucination experienced daily by billions', and as the abstraction of the data in the 'banks of every computer in the human system'.[167] A similar definition was also presented in Gibson's later book *Mona Lisa Overdrive*, wherein cyberspace was conceptualized as 'all the data in the world stacked up like one big neon city, so you could cruise around and have a kind of grip on it'.[168]

As Gibson's descriptions neatly imply, cyberspace is not a digital, electronic phenomenon *per se*. Rather, it is a shared mental understanding that permits us to 'get a grip' on the enormity and power of the on-line revolution. In turn, this may lead us to question whether cyberspace really is a 'new world' in which to engage in fresh forms of social and economic exchange.

As noted by Allucquère Rosanne Stone, the heroes of Gibson's fictional sagas are cowboys of a new frontier—renegade 'outlaws in a military-industrial fairyland'.[169] Gibson's lead characters also consciously 'park' their bodies in the real world during transportation to the cyberspace domain.[170]

In fact, as well as fiction, the notion of *leaving* the real world to *go into the virtual reality of cyberspace* is becoming an increasingly powerful interface metaphor. As Stone for one argues, 'computer engineers are fascinated by VR not only because they can program it, but because they can inhabit it'.[171]

As already explored in the chapters of **Part I**, the immersive 'visitation' of 'cyberian' virtual realities is destined to become more and more run-of-the-mill. Indeed, as VR guru Jaron Lanier enthuses:

> Just as the fish donned skin to walk the Earth . . . [so] . . . we'll now don cyber suits to walk in Cyberia. In ten years most of our daily operations, occupational, educational, and recreational, will transpire in Cyberia. Each

of us will be linked in thrilling cyber exchanges with many others whom we may never meet in person. Fact-to-face interactions will be reserved for special, intimate, precious, sacramentalized events.[172]

We don't, however, have to don VR headsets, datagloves and their ilk to access a potentially 'alterative universe' on-line. Conceptually at least, many people *already* live in three different realms. The first is the physical *realspace* in which we and all other animals are born as flesh, and in which we may breathe and run and procreate and pollute. The second is the mental landscape of *mindspace*—the first realm to be created within and between human beings in a slow evolution that has come to separate us from other animals. Finally today, there is the *thirdspace* of cyberspace—that digital, 'synthetic physical and mental' realm crafted not in language and culture, but in patterns of electrons that may be used to help conceptualize our experiences as temporary cyborgs in tandem with our machines.

Perhaps with good reason, many eloquent technosceptics are already concerned about the rising influence of the 'digerati' who proclaim the conceptual thirdspace of cyberspace as a bold, new frontier. Mark Slouka, for example, bemoans the risk that in the safe, antiseptic 'non-space beyond the computer screen' we may come to kick the habit of reality entirely, and lose 'all trace of earth upon our shoes'.[173] Indeed, as he goes on to argue:

> Cyberspace may distract us from the job in hand—blurring the boundaries between biology and mechanics. Most of the human race are more immediately interested in survival than transcendence. Whilst we wonder through virtual forests, real ones may burn.[174]

Many others concur with Slouka in his fear that the real significance of our 'retreat from the world' may not be in the technology, 'but in the *attitude* that makes it appealing'.[175] Indeed, as a barrier against harsh realities, the virtuality of cyberspace risks becoming a 'social retreat accompanying a loss of the public and social space of the cities, the aesthetic, sensual, and non-human space of the country, a privatization of physical space and a disembodiment of daily life'.[176]

Our emergent, VR thirdspace may also become a realm in which we will rent graphical bodies, genders and personalities as a 'commodity fetish'.[177] Any mass-entry into cyberspace hence risks disturbing and even destroying our individual sense of identity. Too 'easy and indulgent for our own good', some have even argued that present day, text-based virtual communities are already potential thirdspace domains of 'mind candy'. For example, as Shereen Chang of the *Brainstorms* virtual community recently posted:

> It's negative to forgo sensory reality for a virtual one—I've personally regretted losing sight of priorities and spending time/effort online that wasn't deserved, even with people I thought of as friends. I never fully account for the virtual world being as amorphous as it is.

If only due to the degree to which it is supported or bemoaned, it seems clear that cyberspace *is* now strongly perceived as a new land; as a thirdspace in which humans may increasingly live. This said, any acceptance of cyberspace as a new frontier need not necessarily imply that it is wise to think of on-line interactions as 'separate' from those experiences and interactions that sustain and shape us in our 'non-virtual' lives. As passionately argued by Elizabeth Lewis, another member of the *Brainstorms* virtual community:

> What is 'sensory reality?' Why isn't what we do online part of that? I really don't get the splits between 'real' and 'virtual' life. This is part of what I do in my real life. We don't have TV lives, sports lives, school lives, cooking lives, reading lives—why virtual ones that are so distinct from everything else? The biggest split people made in recent years in their lives was between their work and everything else. Even that was faulty, and people who had too much personality split between work and the rest of their lives paid a price.
>
> I still think we all have this incessant feeling of shame about being online because it's just plain enjoyable, and god forbid we enjoy ourselves . . . .
> I also tend to reject the idea that this is 'mind candy'. As I said elsewhere, some folks like to think of this as theatre or performance, but I suspect they feel that way about the world at large—that all the world's a stage for their

amusement. There are people at the other end of the modem, and I'm interacting with them. That seems like something more.

DIGITAL ADDICTION?

Even when and if cyberspace is accepted as constituting a new human frontier, the question still remains of whether we *ought* to mass-occupy its realm. Some of the arguments presented in the previous section suggest that journeys into cyberspace will socially de-couple us from the 'real world'. Authors like Stone also imply that much of the physical/ biological to symbolic/metaphorical transformation that occurs as we 'penetrate the screen' is inherently 'bad' for the human condition. The risk of addition to life-on-line is also highlighted as a danger of the cyber age by those who believe that human beings ought not to evolve to routinely interact outside of the physical world.

However, looking to the flip-side of the on-line coin, information highway advocates often contend that any pending mass human colonization of cyberspace will *increase* and *enrich* human social and well as business interactions. Many Internet users report new friends made on-line who they later meet 'in real life' or 'IRL'. And when economic and other circumstances force lovers, families and friends to be parted in realspace, increasingly their on-going mindspace links may be sustained through cyberspace interaction.

As the social usage of the telephone has for so long demonstrated, people parted in the physical world have the potential for rich interactions in cyberspace with even the most basic technologies. And the technologies and multi-sensory interfaces of the 21st century won't be basic at all. Bill Gates once famously dated a woman in another city in a relationship sustained by cellular phone and computer. Similarly, romances between university students sparked via an Internet relay chat (IRC) or virtual community exchange are increasingly common.[178]

However, even if reluctantly persuaded that all time spent on-line does not have to be socially or culturally 'bad', many people still worry that accessing cyberspace may become addictive. Indeed, it has been argued that cyberspace is starting to manifest itself as a powerful, mind-numbing drug. As Barry Sherman and Phil Judkins only half-joke, 'you can always tell people who have just emerged from virtual reality: they are the ones who try to walk through walls and cross the room by pointing with their finger'.[179] Perhaps even more worryingly, as Howard Rheingold reports in *The Virtual Community*, some students accessing 'multi user dungeon' or 'MUD' network games have been known to regularly spend seventy or more hours a week on-line. The fact that certain forms of on-line activity may become 'as destructive to your life as chemical dependency' can therefore not be entirely dismissed.[180]

The above noted, we may reasonably question *to what* people on-line actually risk becoming addicted. May would argue that it is computers, networks and other cold technologies. However, as the Internet becomes dominated as much by social as by information and economic exchange, a case may be made that many on-line compulsives are actually becoming 'socially addicted' *to each other*.

Allucquère Rosanne Stone argues that children may soon spend more time playing computer games than watching TV, and that many may see this as bad. However, she goes on to consider the opposite interpretation that 'it is entirely possible that computer-based games will turn out to be the major unacknowledged source of socialization *and* education in industrialized societies'.[181]

Stone also reflects that 'technologies that enable near-instantaneous communication amongst social groups pose old problems in new guises'.[182] More pessimistically, Mark Slouka bemoans how our culture has 'bought into the New Age lock, stock and modem'.[183] However, taking a broader, more positive stance, a whole host of other wider visionaries choose to perceive any 'addictive shift' of human beings from realspace to cyberspace as inevitable, essential, and even beautiful.

Taking a largely fatalist viewpoint, Stewart Brand states that 'technology marches on, over you or through you, take your pick'.[184] He then comments that 'the wired world is a teenager with a new car, taking dumb risks, finding new freedoms', and that to watch such self-discovery is a privilege, if also gruelling and sometimes tragic.[185] More fundamentally, Myron Krueger comments that 'the human–computer interface is [now] a permanent part of the human condition'.[186] Virtual community gurus John Hagel and Arthur Armstrong also positively comment that 'networks are ultimately about the connections they create between people'.[187]

However, it is veteran technophilosopher Timothy Leary who puts the strongest and most passionate case. As he explains in *Chaos & Cyberculture*, 'discovering cyberspace' will, for many, be like discovering sex in adolescence. Suddenly new unused circuits in the 'info-wired organs of our brains' will be awakened. In turn, these will extend our minds, enrich the use of our physical bodies, and allow us to find more compatible 'brain mates'.[188] As he further enthuses:

> Intimacy at the digital level programs and enriches exchanges in the warm levels. You do not lessen the richness of your murmur-touch-contact with your lover because you can also communicate by phone, fax, and hand-scrawled notes. Warm-breath interactions with your touch-friends will be more elegant and pleasant with the digital-reality option added.[189]

Savoured or feared—and whether or not they become destructively addictive—lives part-lived in cyberspace loom almost inevitably before those of us who don't want to hurt ourselves, our quality or life, or our chances of 21st century survival.[190] We have become as 'addicted' to computer networks in the late 20th century as we became to electricity in its early decades, and to automobiles in its middle and older age. What's more, our race's colonization of cyberspace is likely to prove far less detrimental than our growing mass dependence on the car.

Rather than debating the dangers of digital addition, technosceptics and policy makers would therefore be better to address the economic, cultural and social divide that is already opening up between the wired and non-wired citizens of our single planet. As Nicholas Negroponte warns in *Being Digital*, there is a very real danger that 'an entire sector of the population will be or feel disenfranchised'.[191]

Clearly the fact 'that it is technically possible for information to be available to everyone at little cost in no way assures that it will be'.[192] As a result, having the means to become 'cyber dependent' may soon determine on which side of the poverty divide most individuals and nations will find themselves. Or as Sherman and Judkins starkly predict, it may be *information famines*, as well as droughts and poor crop yields, that will disadvantage so many people in many African and some Far Eastern countries in the decades ahead.[193]

Large numbers of individuals will probably continue to complain that digital addiction is a curse of the modern age that will force us into a 'robotic lockstep' with computers.[194] However, an even greater percentage of the population are likely to wish that they had the *choice* to become dependent upon daily forays into the cyber-realm.

FRICTIONLESS IMMEDIACY?

Now that the roads linking realspace, mindspace and cyberspace have been opened, it is unlikely in the extreme that they can ever be barricaded shut. How individuals and organizations learn to cope with the immediacy of on-line interaction—and how in turn this impacts upon their saving or wastage of time—is therefore destined to become a very big value creation issue indeed.

As noted by Pattie Maes, the Associate Professor who runs the Software Agents Group at the MIT Media Lab, what will be most valued in future will *be* time. Value creation in the future will consequently depend upon 'services [and] products that help save time'. Or as Douglas Rushkoff similarly noted in part-response to my question of 'what will be most valued in future?':

> People are coming to value time itself—time for contemplation, and time for non-intermediated contact with other human beings.

The above suggests that we perhaps ought to stop focusing our economics quite so strongly upon land, labour, capital, or even the dreaded 'new' productive resource of information. Rather, in a world of near-limitless technological possibility, time itself may be destined to become the most valued economic and social input of the 21st century.

As already noted in **chapter 2**, Bill Gates sees the 'I-Way' of the Internet as a pending 'low-overhead shopper's heaven' in which truly 'frictionless capitalism' may save people a great deal of time as well as money.[195] According to Gates, by buying on-line, individuals will be able to save the 'downtime' frictions of travelling to shops, standing in queues, and travelling home again. Rather than stepping out under a real sky in real rain, it will be possible to make purchases with the click of a mouse pointer, the press of a digit on an interactive TV remote control, or the gesture of a hand in a virtual reality mall.

Already Levi manufactures jeans for individual customers who supply their measurements over the Internet. New PCs may also be 'built' to customer specification on a web page, whilst a myriad of shopping malls, estate agencies, banks and media stores can now be visited on-line. In aggregation, as Internet selling grows, so many customers will come to perceive the travelling and waiting to be endured in many other shopping channels as an unacceptable cost. Granted, many people may like surfing *real* highstreets and malls. However, as middleware rises to replace the buck-taking middleman, customized, cheaper products may quickly prove to hold an equal attraction. Whilst only twenty-seven per cent of women are happy with the fit of jeans bought off-the-peg, Levi can now guarantee satisfaction from denim product custom-ordered over the Net.[196]

Even Bill Gates, however, pauses to point out that technology may magnify both efficiency *and inefficiency*.[197] Once we know what we want—for some already a tall order—we may, in the near future, be able to get it very quickly just that way via an on-line interface. However, the accompanying quickened pace of consumption may simply lead many people to purchase more, discard more, and value less. Also, whilst Bill Gates may note that he is 'so much more limited by time than money',[198] this is not the case for the majority. As a result, consumption that is too easy may, for many people, turn into a debt-mounting curse rather than a time-saving blessing.

This point noted, we also need to remember that the near-instant nature of boundryless, on-line interaction will not just transform markets and their economics. In addition, the organizational and social structures that govern our wider lives are likely to change. As Douglas Rushkoff notes 'the dollar oversimplifies the complexities of a working society'.[199] In particular, people who get used to labouring in groupware-mediated 'virtual teams' may not necessarily want to relinquish their control over such time-efficient technologies at the end of their working day. Rather, they are likely—at least on some occasions—to want to employ them to equal benefit in other spheres of their lives. As Timothy Leary once argued:

> People who use cybertechnology to make fast decisions on their jobs are not going to go home and passively let aging, closed-minded white, male politicians make decisions about their lives. The emergence of this new open-minded caste in different countries around the world is the central historical issue of the last forty years.[200]

Governments, social forums and community leaders—in addition to managers charged with meeting customer and shareholder needs in global markets—may not be able to ignore the challenge of moving their activities into mass-interactive real time. As will be explored in the next chapter, this may particularly be the case when dealing with the desires and demands of the younger generation. Although as Mark Slouka warns, 'what's good for business is not necessarily good for culture'.[201]

I would, however, suggest that on an individual, social level, there can be little that is detrimental to the human condition in the ability to engage in broad one-to-one and one-to-many immediate and frictionless interactions. Certainly, in my experience as an early resident of the Internet, such communications prove enriching far more often than they are occasionally mentally detrimental.

To cite just one example, last December I received an out-of-the-blue e-mail from somebody who had read the prequel to this book—*Challenging Reality*—on a train from Singapore to Kuala Lumpur. Whilst it is always nice to get such electronic 'fan mail', I found the most stimulating part of the message the way it was signed-off as follows:

> If not for the Internet, I doubt I would write this message on a sheet of paper, insert it in an envelope, put the contact address on it, stamp it, and throw it to the PO.

For me at least, this example above powerfully captures the potential *social* frictionless immediacy of the Internet. I freely accept that, if we let it, the Internet may eat into far too much of our precious time. Yet it can also enrich our scant years by allowing so many distant minds to touch, and in the process to shrink the planet. We have started to get used to electronic immediacy and reduced economic frictions in both the business world and consumer markets. However, I suspect that the greater human pleasure of less constrained and more immediate *social* exchange remains, for many, an unrecognised benefit of on-line inhabitation that has still to be broadly reaped.

THE ALTERNATIVE TO CONSUMPTION?

Back in **chapter 3** I suggested that, by turning facilitated human relationships and community affiliations into a market commodity, we may soon witness the emergence of a *gentler mode of capitalism*. In tomorrow's potentially relationship-rich but consumption-poorer markets, capitalist structures may once again prove an effective mechanism for a mass-improvement in many people's quality of life.

Even a slight economic switch away from physical trade, and towards the sharing of group relationships and affiliation, can only lessen the detrimental impact of the present generation upon the human habitat of tomorrow. Indeed, every dollar spent on fostering a lifestyle of *interdependence over independence* may mean just slightly less pollution and a more plentiful physical resource supply for both ourselves and those to come.

Other voices from cyberspace appear to agree with my view that, in a world of mass and immediate connectivity, our attitude to independent, individual product aggregation may soon change. Most powerfully, as Douglas Rushkoff noted when I asked him about future value and its creation:

> At the end of a marketing and consumption boom, our society is coming to realize just how hollow the pleasures of consumption and acquisition really are. The instantaneous gratification made possible by Internet-buying will accelerate this process of disenchantment with acquiring stuff and aspiring to acquire stuff. We will continue to reckon with the fact that our only truly fixed resource is time, and value the time we spend doing what we really want—which will not be data entry.

In a digital society, value comes from an object or person's genuine, physical connection to something real. First edition signed books, officially Beanie Babies, special edition baseball cards. An object's value will be directly proportional to how much time and care another human being put into it.

Cyberspace is perhaps destined to remain a little too 'unreal' and clinically antiseptic for some people's tastes. However, it may yet prove a more socially-responsible and planet-friendly environment than realspace in which to spend at least *some* of our work and leisure time. Our challenge, then, is to find the most effective (and not just efficient) way in which to weave cyberspace interactions into our organizational and domestic lives.

Bill Gates *et al* will no doubt continue to argue that the information highway will magnify the 'clear advantages that capitalism has already demonstrated over other economic systems', and that customers everywhere will enjoy the benefits.[202] In many important senses, supporters of this theory are also likely to be proved correct. However, there remain subtle and significant human and cultural undercurrents to navigate in shaping futures with technological tools that few—and maybe none of us —yet fully understand. For as Brenda Laurel reminds us in *Computers as Theatre,* the primary concern with new technologies like virtual reality 'is not constructing a better illusion of the world; it is learning to think better about the world, and about ourselves'.[203]

A LIFE OF ITS OWN?

In a recent magazine interview, Mira Furlan—the actress who plays Ambassador Delenn in *Babylon 5*—spoke of her 'fear' of the Internet. As she explained:

> The Internet is a scary thing, because anything that you said or somebody said about you, it has this incredible life that you can't influence, and it can get out of hand.[204]

The above concern is hardly atypical. Mark Slouka for one fears the emergence of a 'digital hive' in which interconnected human beings come to exist as cyborg players within a global network machine. Nicholas Negroponte more neutrally agrees that 'when a delivery system that looks like the Internet is used in the general world of entertainment, the planet becomes a single media machine'.[205] Stewart Brand also describes how 'a global computer is taking shape, and we're all connected to it'.[206] Similarly Douglas Rushkoff muses that 'the creation of the datasphere marks a hardwiring of the planet itself'.[207] Or as I myself noted in *Cyber Business*, a single 'global hardware platform' is now emerging across which a great deal of organizational, economic, social and cultural activity is increasingly destined to flow.[208]

Whether our emerging, technological interconnectedness is 'good' or 'bad' largely depends upon whether one chooses to focus upon its organizational, economic, social, cultural or broader political implications. In an organizational context, global connectivity today empowers the creation of global information infrastructures without which global firms could not possibly operate in real-time. Similarly, even across small geographic regions, groupware systems used to interconnect human, technological and financial resources into 'virtual organizations' and 'virtual teams' have become a productivity enabler without which many firms could not survive.

Whilst similarly considering the organizational and economic drivers and implications of an electronic, borderless world, Don Tapscott firmly states his belief that:

> The Age of Networked Intelligence is an age of promise. It is not simply about the networking of technology but about the networking of humans through technology. It is not an age of smart machines but of humans who through networks can combine their intelligence, knowledge, and creativity for breakthroughs in the creation of wealth and social development.[209]

Tapscott goes on to put forward the increasingly common argument that an internetworked level of human consciousness may one day emerge. Such a 'promise fulfilled' form of 'networked intelligence' may even be extended from individuals to organizations. It may hence provide that 'missing link' needed in the realization of true learning organizations,[210] as previously discussed in the last chapter.

Socially and culturally, however, there remain many who fear humanity being market-swamped by the living, digital homogeneity of a blanket, all-pervasive planetary technoculture. The potentially electronic interlinkage of *every* human being and smart machine on the fortunate side of the poverty divide may lead the wired majority to champion interdependence over independence. It may also result in us taking more care of our physical environment. However, as Allucquère Rosanne Stone argues, in the process of global connectivity a 'deep restructuring' is being triggered in the boundaries between 'nature' and 'technology'.[211] Or as Stewart Brand also reflects as he ponders the emergence of the 'cyborg civilization' or 'cognitive planet' that may soon follow the networking of today's 'global cash register':

> Global consciousness is not everybody's idea of a good thing . . . . there is the threat of the global jukebox and the global movie projector weakening cultural identities worldwide. [What's more] . . . the means of electronic communication defence have to be invented while the damage is being done, and all the skilled inventors work for the invaders.[212]

Academics Kevin Robins and Les Levidow appear even more concerned. As they note with some passion:

> Virtual culture raises questions about who and what we are. It encourages us to see ourselves as if we were cybernetic organisms—confusing the technological and organic, the inner and outer realms, simulation and reality, freedom and control . . . . Through a paranoid rationality, expressed in the machine-like self, we combine an omnipotent phantasy of self-control with a fear and aggression directed against the emotional and bodily limitations of mere mortals. Through a regression to a phantasy of infantile omnipotence, we deny our dependency upon nature, upon our own nature, upon the "bloody mess" of organic nature.[213]

Mark Slouka simply states that he would like to 'set up a good roadblock' before the digerati and their digital hive threaten individual abilities to 'feel compassion, loyalty, and love for others'.[214] However, it ought to be more broadly noted that any such 'threats' are not just to individuals alone. Indeed, perhaps destined to have a more obvious impact on our lives is the way in which cyberspace may become the dominant structural overlay of mass civilization.

For two centuries or more, modern nation states have been the major sub-planetary frameworks governing global resource allocation, not to mention cultural diversity and divide. However, as many commentators now claim, any global tapestry woven from nations is becoming increasingly threadbare. Not least this is because nations often combine resources at the wrong level of economic and cultural aggregation. Today's nations founded in realspace are usually too large to be shaped by a single, meaningful economic or cultural homogeneity. At the same time, they are also usually too small to muster significant economic or cultural leverage on a global scale.

Kenichi Ohmae of McKinsey & Company argues that economics, not national politics, will increasingly define the landscape against which all else must operate.[215] And has already been argued, economics and new forms of business organization go hand-in-hand with the sprawling advance of Slouka's much-feared digital hive.

Many of us with access to technology may soon want or need to become citizens of a global, cybernetic entity. However, this need not imply that we will lose parallel affiliations with other valued, large-scale structures. However, what is increasingly unlikely is that these frameworks will be geographically-insular nations and their governmental economic regimes alone. In one hundred years time we may well look back to the present age as that time in which the machines of governments were superseded by the global machine of the Net. In the medium- and long-term, humans will become far less *citizens* of nations, and far more *members* of on-line virtual communities, loyal *subscribers* to certain Internet service providers, *users* of a particular digital currency of first choice, and *affiliates* to an adopted security blanket of a private welfare state.

Such wide-ranging developments may not all spring from electronic, global connectivity and the inhabitation of its common cyberspace alone. However, the Internet (or whatever form of network and interface it evolves into) is likely to provide the future glue that will hold together many strands of human lives, cultures, and governance regimes.

With little doubt—and again, love it or loathe it—the Internet has become a global, technological *entity*. In the sense that it adapts rapidly to survive, it has also begun to take on a life of its own. However, it remains a life totally dependent upon a close, symbiotic relationship with a human race that increasingly draws value from mass interdependence. For those who harbour concerns that so much is under threat as a digital hive evolves, one may therefore at least offer the comfort that the machine cannot live without woman and man. Yet human beings can, just about, still survive without the learning child of their networked planetary machine.

* * *

A MEDIUM IN BLACK-&-WHITE?

In 1939 at the World Fair in New York, shoppers with between $200 and $600 to spare could have purchased one of the very first television sets to go on sale in the United States. Such technological marvels boasted black-and-white tubes a mere five inches in size. Perhaps hardly surprisingly, many people at the time therefore quickly dismissed television as a fad that would never catch on. After all, as early sceptics argued, how many people would be prepared to spend large proportions of their time gazing at a tiny, fuzzy, black-and-white screen?

I mention the above in the firm belief that, in comparison to a modern television set, the Internet today is barely in five-inch fuzzy black-and-white. Today, most people are not on-line. Most also perceive cyberspace as a realm to be accessed at a desk via a screen and keyboard upon which the majority remain unable to type. However, to associate the future of the Internet with two-dimensional screens, text, graphics windows, and keyboards, is to make as big a mistake as those blinkered individuals who dismissed television because its early black-and-white images were only five inches across.

As discussed in earlier chapters, virtual reality interfaces, voice control, and smart software agents, will over the next couple of decades transform the Internet into a rich, broad-access *experience metamedium*. No longer will people have to use flat screens and keyboards unless they want to. By spoken command, hand gesture, virtual world immersion, or even thought, Net access will become just as common from smart buildings and worn devices as it will be from desktop workstations and pockets.

However, even before ubiquitous cyberspace access becomes a reality, we will hopefully stop making the mistake of solely associating the Internet with personal computing. Rather, the Net—the thirdspace of cyberspace—will come to be respected as that common arena in which many people work, trade, socialize, relax, and hence part-live. And rest assured I do only mean *part*-live.

Few seem to doubt that the cyberlands of the Internet are a new human frontier. Many fear the consequences of a 'technological addiction' to this realm as Marshal McLucan's 'global village' becomes the largest—and perhaps only—country in the world. However, most on-line 'addicts' in the future are as likely to become dependent upon geographically distant *other people*, as they are on digital technology *per se*.

Fears of a 'faster world' of point-and-click immediacy are today well founded and unlikely to be quelled by those technological, social and economic developments ahead. However, we ought perhaps to remember that as individuals we all fuel the engine of economics that is in turn now causing much of affluent humanity to shift gear into the fast lane. There is also, perhaps, the possibility that more *human* time—both on- and off-line—may be both valued and afforded in future if our capitalist structures become more focused upon relationships and their maintenance, rather than on physical consumption alone.

Ultimately, whether the broad interlinkage of ourselves and our machines through cyberspace will be good or bad for both individuals and their societies remains a difficult judgement to call. Clearly there *are* dangers to individual and cultural liberty in a single metamedia that enshrines the planet. However, the Net looks equally likely to remain a many-to-many interactive media that will continue to be able to champion individual expression over corporate or political oppression. Certainly no state or organization has yet succeeded in censoring the Internet. And that comfort alone makes its global network of interconnected human beings historically unique.

7
The neXt Generation?

ACCORDING TO A RECENT Intel advertising campaign, when they 'show people what the Pentium II processor can do for a PC, they want to get it'. Except, that is, for kids—who 'already get it'.

Intel's advertisement is clearly intended to imply is that there is something different about the attitude of young people to new technology. There probably is. Though I don't think that Intel knocked it exactly on the head with their dancing-bunnymen TV extravaganzas. The 'kids get it', certainly. But probably more than that, the kids *don't need to get it*.

Desktop computers first became available from the likes of Tandy Radio Shack and Apple in the late 1970s. The IBM PC was then launched in 1981. Today's toddlers, children, teenagers, and even twenty-somethings, may need to be taught *about* computers. However, they will never need to learn *of* them. Nor will they ever need to learn to *accept computers* as new components of their life order. To most people below the age of thirty who live in an industrialized nation, desktop computers are no more 'new' or 'revolutionary' than televisions, radios, washing machines, cars, electricity, supermarkets or McDonalds.

Computers and their internetworked, global culture have always been part of the landscape of today's youth. As a consequence, to most young people computing is *transparent*. Young people simply don't see the screen, mouse and keyboard. Rather, they tend only to perceive the information, work or game that today happens to be presented on a monitor. Unlike many of their thirty- or forty-something-plus parents, most young people also don't tend to get excited about a particular PC's peeling-sticker speed rating. Indeed, it is usually only when the software application before them is running unacceptably slowly that most young people ever bother to comment upon the speed or other hardware characteristics of a particular computer.

In terms of valueware, many older people still have a lot to learn from the younger generation's transparent attitude towards computing. Sorry Intel, but the kids don't care if there's a Pentium II processor in their PC (except, perhaps, as a symbol of global brand affiliation). Rather, the kids care only what any machine—fast or slow, cheap or expensive—can do for them. The kids get it to such a large degree that the kids *don't need* to get it.

THE FUTURE WALKS AMONGST US

This final chapter of value perspectives is all about the defining characteristics of tomorrow's workers, customers, innovators, and product champions. Fairly obviously, 'Generation X'—those people born between 1963 and 1981—will have a major and increasing impact upon value creation in future markets and organizations. Indeed, already around twenty per cent of all employees in major Western companies are 'Xers'. Learning to understand the *neXt generational psyche* of this more and more influential group of people is therefore of paramount importance for all managers and marketeers.

Even more broadly, it is also important to recognise how the *values* of the first-ever information society generation will increasingly influence workers and customers in other age bands. As just one example, some Internet service providers are already experiencing a boom in the growing number of old-age pensioners—or 'Saga surfers'—who are getting wired to browse the web and to exchange e-mail with their grandchildren.

As Douglas Rushkoff argues in *Children of Chaos*, today's kids are our test sample—our 'advance scouts'—for learning to cope with the Wired Age.[216] They are the 'latest model of human being, and are equipped with a whole new lot of features'.[217] As Rushkoff continues:

> Today's 'screenager'—the child born into a culture mediated by the television and the computer—is interacting with the world in at least as

dramatically altered a fashion from his grandfather as the first sighted creature did from his blind ancestor, or a winged one from his earthbound forebears. Human beings have evolved significantly within a single creature's lifespan, and this intensity of evolutionary change shows no signs of slowing down.[218]

One of the best ways to get a grip on the markets and organizations of the future may therefore be to study intensely the ways in which Generation X now surfs and navigates and *thrives upon* our new, global culture of internetworked, interactive media. As Geoff Mulgan similarly argues in his study of the 'connexity' of humanity's increasing independence *and interdependence*:

> . . . ideas of freedom, the freedom to choose, to exit, to shape your own identity, and to be empowered . . . constitute the ideology of the first generations to grow up against a backdrop of connexity, the roughly two billion teenagers in the world by 2001, a large minority of whom will be familiar with computers, virtual reality and video links, and a common, if highly fractured, culture. Theirs is an ideology that favours tolerance, pluralism and diversity, and that defines the self as a choosing, self-determining entity, not subject to fate, or blind destiny, but rather able to make its own life, and even its own morality.[219]

Many people—and in particular many 'hippy-turned-yuppie' Baby Boomers—have in recent years labelled Generation Xers as 'slackers' who exhibit little commitment or concentration, and who hence end up in low-status, low-paid 'McJobs'. Indeed, the term 'Generation X' comes from the title of a cartoon strip and then 1991 novel by Douglas Coupland about a group of jaded twenty-somethings who had dropped out of the rat race. The 'lost generation' to follow the Baby Boomers who had 'devoured all the good stuff' was also so-labelled as 'X' by *Time Magazine, Newsweek, BusinessWeek*, and *Fortune* in 1993.[220]

However, as researcher Bruce Tulgan has recently shown in an extensive survey, the popular media stereotype of Xers as an arrogant, slacking generation of 'lessness' is a misconception. Xers may be 'galaxies apart' from the post-war Baby Boom generation who were 'forced' to grow up from flower-power idealism to 1980s greed and

1990s realism. Yet, as Tulgan claims, this does not mean that Xers can simply be dismissed with a blanket, derogatory 'slacker' label. For example, Xers may be 'self-builders' in the workplace who appear less committed than Boomers to climbing corporate ladders. However, this is simply because long-term organizational relationships can no longer be trusted.[221] Xers are a committed generation, but largely to careers and professions, rather than to individual organizations.

Xers also thrive on immediacy and information. They hence require a management style that recognises the former and provides the latter. As Tulgan summarizes, rather than being a 'lost' generation, Xers are intelligent, creative and hardworking.[222] Some Xers have been conditioned to self-reliance by a latch-key childhood.[223] Many also assess successful management relationships in terms of an available richness and depth of belonging, learning, entrepreneurship and security.[224]

Regardless of what other generations may choose to believe about them, Xers *are* the human force that will drive all of our futures. Some Xers are already computer programmers and social activists whose actions significantly influence contemporary society and its products, services and organizations. The rest have yet to enter, or to make an impact upon, the job market. However, all Xers are destined to become both future workers and customers. Xer's values—and largely Xer's values alone—will therefore be those that Future Organizations will most need to serve if they are to remain in business. Indeed, as the 'metamedia' that Xers fluently surf continue to spread, so *neXt Generational* values are increasingly likely to become linchpins of mainstream human culture.

For better or for worse, there can be no doubt that the future dominant citizens of this planet already walk amongst us. Organizations founded in a different social and technological era therefore need to embrace the values of this generation, or else risk extinction as cultural evolution marches over them regardless.

As an aid to the above, the following sections explore seven potential characteristics of the *neXt Generation*. Specifically, they seek to explain how future workers, innovators and customers are likely to be differentiated from their counterparts of today in terms of geography, politics and the environment, shared experiences, loyalties, structures, interfaces and culture.

THE MOBILE GENERATION

The *neXt Generation* will know more geographic freedom than any other. Indeed, the majority of people—young and old—across most nations even today have more potential freedom to be mobile than ever before. The reason for this is two-fold. Firstly, transportation costs are falling in relative terms. Secondly, there is an increasing world-wide availability of transport infrastructures and vehicles ranging from personal automobiles to international airliners.

As explored in depth within Tsugio Makimoto and David Manners' book *Digital Nomad*, the human race is increasingly on the move. In the late 1990s, three and a half million people daily took a scheduled air flight, whilst at any one time there were some 300,000 people flying over the United States. In the Western world, passenger air travel continues to grow at around five or six per cent every year—an impressive figure until one learns that air traffic in Asia is expanding at around twenty per cent per annum.[225]

Literally *billions* of people are also carried on rail journeys every year, whilst the world's population of six hundred million automobiles is growing by thirty million per annum. Even today there is a car for every 8.6 people on the planet, and—to the alarm of many—humanity's insatiable appetite for more and more personal geographic freedom only continues to grow.[226]

Reeling from the above statistics, it is perhaps not difficult to accept Makimoto and Manners' claim that 'in a decade most people in the developed world will be free to live where they want and travel as much as they want'.[227] However, the unprecedented geographic freedom of the *neXt Generation* of customers and workers will not be dependent upon 'improved' transportation provision alone. In parallel, the technologies of cyberspace will afford individuals the freedom to travel with none of the traditional inconveniences of a nomadic existence.

Increasingly, travel need not be accompanied by any dislocation from information, work and entertainment. Nor need future travellers lose touch with family and friends. As Makimoto and Manners contend, a 'complete nomadic toolset' offering computer power, information storage, language translation, digital money, and wireless global communications, will soon become a *technological* possibility. Costing around $500 or less, and weighing no more than two pounds, such a personal device will offer people the facilities of their homes and offices in their pocket. Future workers and customers will subsequently have the choice to free themselves geographically from fixed homes and workplaces. They will therefore enjoy the option to roam or to settle as and when and where they please.

At the other end of the scale, some members of the *neXt Generation* may instead choose to use cybertechnologies to enable them to 'cocoon' in one geographic location from which they will rarely, if ever, roam. Given the common human lust for travel and exploration, 'static individuals' will probably remain in the minority. However, it remains important to appreciate that for both roamers and cocoonists, emerging technologies will make geographic mobility more and more of a personal *choice*.

The 'geographic span' of many future workers and customers also looks destined to increase whether or not they choose to be digital nomads or cocoon-dwellers. As explored in the last chapter, the Internet is already seeding the dawn of a single, global culture in the wake of a world previously divided by political barriers that held ideologies as well as people apart. Global marketplaces accessible to individual as well as corporate consumers are also opening up as buying and selling on-line become more common.

A highly mobile generation with more and more geographic choice is likely to have profound implications for many markets and organizations. Not least, the multi-billion dollar travel industry—already employing one in nine workers—will probably continue to expand.[228] However, industries such as financial services may also be affected if customers —and in particular those young and old customers with the greatest opportunities to roam—spend more on travel and less on fixed homes and mortgages. For governments, deciding how and where to tax citizens whose home location may simply be 'the planet' will undoubtedly also become a major headache.

Pinning down under which legislative systems many future products and services will be regulated will similarly prove an increasing problem. Many companies entering global markets will also need to broaden their marketing techniques and strategies in order to embrace the global Net. Indeed, unless they adopt new 'wired' marketing mindsets and technologies, many companies will find it all but impossible to maintain ongoing relationships with an increasingly mobile customer base.

Finally, for some of tomorrow's workers and customers, the conceptual enlargement of geography itself may influence future concepts of their 'home' and 'work' location. Whilst realspace may remain finite until off-world travel becomes more common in the mid-21st century,[229] the 'geography' of cyberspace will probably continue to expand. Roaming or cocooned in landscapes broader than most previous generations can have dreamt to know, the planetary citizens of the future will present many organizations with tremendous marketing and product/service supply challenges and opportunities indeed.

THE GLOBAL GENERATION

Increasing mobilities in both realspace and cyberspace are also already impacting upon the *neXt Generation*'s appreciation of *politics and the environment*. Today's young people are the first to grow up with real-time global news, not to mention the shared concerns and transnational culture of Greenpeace, CND and MTV.[230] To an increasing proportion of the population, the Cold War is not just the past, it's a part of history. The *neXt Generation* are indeed the first truly planetary citizens, not just in terms of mobility, but also as a result of their moral and political appreciation of humanity's global *inter*dependence.

Common-citizenship concerns for nuclear disarmament and a greater care of Mother Earth, coupled with the global immediacy of the technologies of cyberspace, have already significantly eroded the perceived relevance of nations and national politics for many young people. Commentators around the world note how Generation X is less likely than any other demographic group in history to *register* to vote, let alone to partake in a political election.

As Bruce Tulgan explains, many Xers don't see politics as being 'worth [their] time and energy'—much less a 'center-stage for expressing [their] self-hood'.[231] This may partly be because Western politics in recent decades has come to be linked with sleaze, surrealism, ineffectiveness and broken promises. However, more broadly, the disenfranchisement of many young people from national governments probably also reflects the fact that politics has become out of touch with many of the issues that impact upon individual futures and lives.

Governments have, almost by definition, to foster an *international* rather than a *global* mindset. The difference? Only that anybody who talks of 'international politics' or 'international business' remains conceptually wedded to a planet segmented *into nations*. Internationalism and globalism—in terms of politics, economics, marketing, culture or trade—remain light years apart. The *neXt Generation*—and in particular many Internet users—know this. To their and our detriment, many political and economic bastions do not.

For the *neXt Generation*, global technologies fuel global cultures through global fears. Rather than the Cold War political hysteria of nation fearing nation, the young of today are more likely to fear AIDS, global warming, social decay, and the implications for local and global societies of an increasing underclass–overclass divide. Indeed, an appreciation of the fungibility of planet Earth can escape nobody who today attempts to think and plan ahead with an open mind.

To most successfully serve their customers, workers, and hence themselves, Future Organizations need to develop a strategic awareness of the rising planetary psyche of a global interdependence fuelled by common fears. Successful future global products and brands are likely to be those with solid, planet-friendly credentials. Exploiting a workforce in a distant land for profit in another country is also less and less likely to reap any long-term business reward. For the *neXt Generation*, the artificial maintenance of national boundaries makes little sense environ-mentally, economically, culturally, morally or socially. For successful Future Organizations, the same will therefore have to be publicly seen and proven to be the case.

THE METAMEDIA GENERATION

Perhaps the most startlingly-different characteristic of the *neXt Generation* is the way in which it has learnt to share *experiences* in new ways by conquering the passive linearity of traditional one-to-many media. With music, video and a variety of text works increasingly available in random-access formats—including CDs, DVDs, and hyperlinked media[232]—so tomorrow's customers, workers and innovators are evolving into non-linear information surfers. As a result, many young people appear to have no significant attention span. Or as the cry so often goes out, 'kids today just don't seem able to concentrate'.

This common belief is, however, a nonsense. The attention span of most young people—and there will always be an unfortunate few delinquents or drop-outs—remains terrific. However, what has changed is the way in which a broad, novelty-seeking *attention breadth* is now more characteristic of many young people than a traditional and linear *attention depth*.

To understand how and why such a transformation has occurred, we need to appreciate how the *neXt Generation* already engage in transmedia communications involving multimedia cross-integrated to such an extent that they effectively comprise emergent *metamedia*. As defined in **Challenging Reality**, 'metamedia' involve the combination of many previously distinct media forms into one synergistic whole. Metamedia may thereby enable the transmission of ideas and identities via the replication, storage and communication of synthetic human *experiences*. In contrast, previous multimedia—such as television or the movies—may only ever prove effective in the communication of knowledge, information or data via relatively few channels.[233]

Because metamedia experiences are starting to be perceived and delivered via a synergistic *combination* of so many traditional single and multimedia channels, they are also beginning to alter the culture and mode of some forms of human communication. Reflecting this, the *neXt Generation*—and children in particular—are becoming surfers of soundbites, rather than passive hoarders and unquestioning recipients of linear information. As Douglas Rushkoff notes, the new generation turns to the media for questions, not answers.[234] Indeed, Xers seek 'truth' and 'certainty' by browsing and comparing multiple media sources, rather than choosing to absorb passively just one or two linear information streams from start to end. Xers are just as likely to sit down to 'watch TV' or to 'surf the web' as they are to settle before a particular, pre-selected single programme or Internet site.

The keys of the handheld TV remote control, and the hyperlinks of the world-wide web, have helped to foster the emergent surfing mentality.

And this may well be no bad thing. Tomorrow's most valued workers, not to mention tomorrow's most info-savvy customers, are likely to be those individuals capable of spotting and creating powerful metamedia experience *patterns* from their non-linear surfing of the amazing content deluge of the information age. To avoid drowning in an ocean of information overload, one has to learn to surf its waves.

Being a collage artist who revels in cut-and-paste will be more important for most of today's teenagers than the ability to sit through and study a passive, two hour, single-media lecture or television programme. As Rushkoff argues, traditional long attention spans teach people what to think.[235] Effectively, they program viewers or readers to ride with a message that may, perhaps dangerously, blinker them into becoming too engrossed, sidetracked, or even brainwashed, with what a particular media baron wants them to believe.

The broader implications of the rising evolution of multimedia into metamedia are discussed in more depth in the next chapter. For organizations as both employers and marketeers, such an evolution can also not be ignored—however weird or scary the concept may at first appear. Kids raised on MTV, 'channel-zapping', interactive CD-ROMs, and the Internet, have already learnt how the texture and *feel* of an information source can be more important that its total visual or other content. As a result, they don't necessarily always need to watch, hear or read *all* of something in order 'to get what it's about'. As Rushkoff analogizes:

> The art of coping with a high-tech, nonlinear culture is the art of surfing, skateboarding and probably most of all, snowboarding. For where surfing is a negotiation with the dynamic waveforms of nature, and skateboarding is a negotiation with the cityscape, snowboarding is an immersion in designer discontinuity.[236]

THE TRIBAL GENERATION

As already discussed in previous sections, increasingly the loyalties of the *neXt Generation* are not to governments, histories, or geographic localities. Extended families, too, are less common in many young, latchkey lives. Indeed, as Geoff Mulgan argues:

The long period of peace and prosperity that has come since 1945 has steadily eroded the attachment to traditional values, organised religions, authoritarian political ideas and even nationhood, and replaced all with a profound attachment to autonomy, as well as fostering other values like the attachment to authenticity.[237]

As the paradox of Mulgan's book *Connexity* explores, both indepen-dence and interdependence are rising, hand-in-hand, as global social forces. People are choosing and building their own life-anchors, rather than passively accepting the 'accidental' attachments associated with their place of birth or the beliefs of their parents. As already explored in **chapter 3**, in the place of many traditional geographic or workplace ties, human group affinities are increasingly being exhibited to the 'tribes' of global brands promoted via sophisticated metamedia.

For the *neXt Generation*, there is often little, if any, medium- or long-term stability in their lives beyond their own set of choices. Indeed, we may well reasonably question the whole meaning of the word 'belonging' in an age of mass-disposability, flexible work regimes, and so many families boasting a single adult alone. As Bruce Tulgan examines in his study of Generation X:

Growing up in a habitat of rapid change and social atomization, without the backdrop of 1950s and 1960s idealism, Xers have had sparse opportunities to witness or experience enduring affiliations of any kind—social, geographical, religious or political. Our own family structures, and those of our peers, have not been reliable. We are unlikely to have spent our childhood in one community—and even if we did, childhoods marked by the characteristics of suburban diaspora and the evisceration of community centres. For this reason, even our friendship circles have always been in flux, shifting along with forces beyond our control.[238]

Given the above, it is perhaps hardly surprising that the loyalties of the *neXt Generation* are already subject to a more 'immediate dynamic', and that Xers 'have a different way of belonging and developing allegiance' than prior generations.[239] Branded trainers, dance crazes, or even forms of self-modification like tattooing and body piercing, are increasingly the tribal insignia of the young. For example, Douglas Rushkoff describes raves as 'intensely tribal, making use of technology to promote [those] deeply spiritual agendas' that may in previous ages have been communicated through national or religious affinities.[240]

On the streets and across the Internet, young people are creating tribes bound together by new sets of values linked to the 'right' sportswear, club, designer drug, or web forum. Even for older generations, labels as much as products and services increasingly matter. Companies like Richard Branson's Virgin are succeeding by trading in loyalty and *trust* across multiple markets and product ranges in a way that may change economic landscapes forever. Cultures and their norms are now designed and exported. They have become part of an ongoing product or service experience, rather than local and demographic barriers to market entry. As Intel's recent TV campaign signals, it is becoming increasingly important to ensure that target customers 'already get' your product *before* it comes to market.

Across independent yet interdependent human minds, centres of customer influence are also changing. Purchasing agendas are starting to be swayed as much by loyalties to branded supply *channels* as they are to single, branded *products*. In the highstreet, 'loyalty cards' are proliferating in an attempt to 'attach' customers to a supermarket or other retail outlet long-term. Yet at the same time as some organizations are entering new markets with old brands—major UK supermarkets moving into the financial services sector, for example—we are witness to a decline in brand inertia.

As centres of influence become more dynamic, so customer loyalties are equally becoming more fickle. Nike may be the footwear of choice today, but not so tomorrow. Companies are therefore starting to appreciate that the most successful Future Organizations will be those that not only *have* strong, cross-market tribal brands, but which also continue to *build* new ones.

It is also becoming more apparent that customer satisfaction and customer loyalty are not the same thing. The risk to many businesses of suffering a public relations disaster capable of damaging a brand across multiple markets is hence growing. Broad brand affiliations built off-line in the highstreet or mall, or on-line in a virtual community, may apparently foster loyalty from tribes of modern customers who will trust a corporation to supply everything from food and clothing to banking services and travel. However, as a result of the rise in branded meta supply channels, a single food scare, publicity scandal, or poorly handled customer complaint, now holds the power to decimate company sales not just in one market segment, but across a wide 'lifestyle' range of goods and services.

THE NETWORKED GENERATION

Partly driving the rising power of global brands and tribes are the increasingly complex networks now characteristic of so many of the *structures* that govern modern lives. Perhaps most notably, lumbering, inflexible, organizational bureaucracies are being forced by competitive environmental pressures to evolve into more fluid and dynamic networked productive forms.

Across many industries, the majority of static, middle-management-heavy corporate hierarchies have now disappeared. Modern customers will simply no longer sustain such costly and unresponsive organizations. As a result, the *neXt Generation* knows only an economy defined by the disappearance of job security. Careers may therefore no longer be built around a long-term relationship with one employer. Instead, successful 'self-built', or 'portfolio' careers are becoming characterized by a strong, individual network of past and present job *relationships*.

Networks of relationships, as opposed to stable structural frameworks, are also increasingly characteristic of many social as well as organizational lives. Over the past few decades, widening geographic mobility has resulted in large numbers of friends and families who no longer live in close physical proximity. Consequently, many of us increasingly connect with some of those to whom we are close via networked communications links. Indeed, in many lives, the telephone—and even the Internet—have become an essential social glue to hold together a large number of valued human acquaintances.

For the *neXt Generation*, many technologies as well as social and organizational movements are fairly obviously most commonly experienced in a networked context. Personal computers in particular are far more valued if made *interpersonal* via their linkage into the global

Internet. As outlined in **chapter 2**, intranets and extranets are also rapidly becoming the key mechanisms through which many business organizations may manage a wide range of their tasks and broader processes.

In aggregation, the *neXt Generation* are destined to spend large proportions of their lives both as the outlying nodes of organizational networks, and as the cores of individual social webs. Many of those key technologies that they will employ every day will also be networked. In turn, as networked relationships come to be more and more effective and hence *valued*, so those technological, organizational and social *interfaces* expected and championed by the *neXt Generation* will also inevitably evolve.

THE POINT-&-CLICK GENERATION

Networked relationships in today's real-time, 24-hour society will continue to fuel human demands for interface systems that offer a high degree of transactional and social immediacy. As previously noted, many citizens of the *neXt Generation* have already been weaned on hyperlinks. Many also already live in the three, overlapping spheres of realspace, mindspace and cyberspace. Across such a New Age terrain, hierarchy has little meaning, whilst a 'tomorrow will not do' delivery mentality has become firmly inbred.

Youngsters in particular want to *click on* rather than *stare at* an advertisement. They then expect to be able to receive associated products or information *immediately*, rather than two days or even two weeks down the line. Technoliterate young people also know that this is now possible. Those companies that permit 'instant' access to their product and service ranges via point-and-click on-line interfaces are therefore likely to sell most successfully to the *neXt Generation*.

Anybody who works in traditional retail has reason to fear an emerging customer generation who are starting to learn the pleasures of information-rich, low-transaction-cost, point-and-click on-line supply channels. Forecasts for the growth of on-line selling continue to vary considerably. However, predictions suggesting a yearly sales volume over the Internet of at least $100 billion by the year 2000 have been consistently made by many respected market analysts. As argued by Michael Willmott of the

Future Foundation, we should also note that too many respected commentators, and too many major IT suppliers, have been predicting and wedded to an explosion in on-line selling for too long for their visions to prove a mere flash-in-the-pan.[241]

The potential savings in transaction costs to be reaped by those organizations that develop on-line selling and/or customer service are already considerable and cannot be ignored. As recently reported by the American Banking Association, the cost of servicing a customer transaction over the world-wide web is about one cent, compared to twenty-seven cents for an ATM transaction, or well over a dollar for 'human' service in a typical branch.[242] Supply-side pressures, as well as *neXt Generational* customer demands, are therefore extremely likely to fuel a massive increase in the use of point-and-click interfaces for future product as well as information access.

As argued in **chapter 2**, direct electronic links between customers and organizations will become more and more common in an internetworked economy in which middleware is already starting to replace the middle-man. In markets dealing in expensive and complex products in particular, integrated web-based advertising, sales, and customer support point-and-click interfaces are likely to become the norm. It's not only that direct-to-consumer prices are likely to be thirty per cent lower on-line.[243] Just as importantly, the expertise available from a sales assistant in a computer or hi-fi shop in the highstreet is simply not going to be able to match that available from a corporate web site or a dedicated virtual community.

Most customers today may say that they prefer to browse a real highstreet rather than to surf a virtual mall. However, the opportunity to pay a lower price for a product about which they are better informed may for many soon prove a powerful incentive to shop on-line. It should also not be forgotten that the *neXt Generation* will use the point-and-click immediacy of the Internet for social as well as information and trading exchange.

People matter to people when they are spending their money. Indeed, it is for this very reason that many continue to claim that business on-line is doomed from the outset. However, it precisely because people matter to people that on-line virtual communities may soon make business on-line socially (and economically) acceptable.

Via a common point-and-click interface, future virtual communities have the opportunity to populate Internet shopping malls with other customers and window browsers as well as high quality, low cost products and information. The non-wired, street-bustling shoppers of today may be unlikely indeed to accept on-line shopping as the norm. In contrast, those cyber-surfing members of the *neXt Generation* may soon view it as the most natural means of acquiring some—though not all—of those goods, services and human relationships that they both need and desire.

THE WIRED GENERATION

In January of 1998, *Wired* magazine reported that in response to the question 'which would you rather keep, your TV or your PC', seventy-one percent of children chose their PC. To many older people this figure may be somewhat surprising. However, it once again only serves to reinforce Intel's powerful perception that *the kids already get it*.

In 1996 Motorola reported that 16–24 year olds were the highest users of IT products. Eighty-five per cent of survey respondents of all age ranges also thought that children would find it useful to have access to the Internet.[244]

For the young, 'being wired' is already a *cultural* phenomenon relating to new patterns of linkage and relationship not just between cold machines, but also between human beings and machines, and hence between one human being and another. To understand the technologies of the Internet *et al* is relatively simple. However, it is the greater hurdle of creating value within the new human landscape of networked, point-and-click global cohesion that really separates those who are 'wired' from those who have yet to connect with the psyche of the New Age.

The *neXt Generation* are the first occupants of planet Earth to live at ease with the crossing of the Fourth Discontinuity. Indeed, the 'Wired Age' itself may perhaps best be defined as that period of time post the Fourth Discontinuity wherein the commonalities and interdependencies that exist between our ourselves and our machines are fully embraced.

With technological, social and cultural change accelerating at so frantic a pace, it remains difficult to isolate all of those key elements of the *neXt*

Generational mindset that set some people apart as being 'wired'. However, having attempted to isolate bullet-point 'wired' or 'future mindsets' in both *Cyber Business* and *Challenging Reality*, I would suggest that there are at least some fairly definite 'wired ways of thinking' that the *neXt Generation* exhibit, and that others may already now begin to practice.[245]

Firstly, I would propose that wired individuals are those who *look through their hardware*. In other words, they are those people—young or old—who accept the transparency of modern, digital systems, and who hence make sure that they see tasks and processes, not screens, mice, keyboards and *Windows* applications.

Those with a wired mindset are usually also highly aware of the dangers of *incrementalism*. What this means is that they don't place too many barriers in their minds that may blind them to the opportunities and threats presented by *radical* change. In particular, wired individuals are careful to be open-minded when it comes to accepting the likely future potential of IT. Certainly, they have developed the ability to 'think out of the box'. Many managers today will happily plan for the Internet revolution. However, some of the same people still dismiss as 'science fiction' the dawning virtual reality revolution that will soon transform on-line information, business and social exchange. By only accepting incremental IT developments as possible and relevant, such individuals hence risk the strategic futures of their organizations in ways that those who are truly wired do not.

As surfers of emerging metamedia, wired individuals take care to *avoid 'hyperreal' visions* so characteristic of the information age. The *neXt Generation* know that a digital picture is worth a thousand lies, and that as films like *Forrest Gump* have so effectively illustrated, the camera can now lie like a rug. Wired individuals hence always verify the texture and content of important data and information from multiple sources. At the same time, they also harbour a healthy scepticism for that on-line 'fact' that can only ever—in human terms at least—tell half of the story.

Whilst being wary of the hyperreal—of black-and-white news and miracle products—wired individuals also trust themselves to *accept incomplete knowledge*. Born into the information age, the *neXt Generation* have quickly learnt that all the majority of networked systems can ever tell them is how much data available there is that they will never, ever know. In the light of this modern reality, the development of the confidence to make a decision, even when all of the potentially-available 'facts' have not been sifted, has become a critical 'wired' competence.

Finally, thriving in a global technoculture of workplace flexibility and on-line, point-and-click immediacy, wired individuals will almost always *opt for action*. They 'just do it'—no longer trusting promised, long-term paybacks that seldom, in their experience, ever justify the wait. Non-wired individuals and organizations who don't accept incomplete knowledge may procrastinate forever, and in the process fall farther and farther behind. In contrast, the *neXt Generation* act in the certainly that there will never be perfect knowledge around the corner, and that the greatest reason to opt for action *is not to be right, but to be learning*.

* * *

GENERATION neXt?

Looking ahead in any attempt to predict the characteristics of a future generation may be as fraught with danger as any other area of future studies. This said, many common threads can be found running through the observations and research studies of a wide range of commentators. It is therefore unlikely that the general thrust of those value perspectives presented within this chapter will be proved wrong.

For better or for worse, the *neXt Generation* look less likely than any other to be passively programmed or 'brainwashed' into accepting those traditions and loyalties that governed the lives of so many of their parents and grandparents. Granted, youth has always been characterised by rebellion. However, whereas the hippies 'grew up' to wear suits and to suck economies dry in the 1980s, it seems far less likely that the young people of today will have to break through a similar pain barrier.

For the first time, young people have become a globally homogenous group. They share global concerns, an emerging technoculture, and instantaneous, global communications. Problems clearly lie ahead for the human race. Yet thankfully, it appears that the generations of the future will face those many crises ahead as one. Some nations—France being a notable example—may continue to bemoan the fact that a homogenous, global technoculture runs the risk of swamping national and ethnic cultural diversity. However, by enabling millions of minds to touch daily around the planet, a global technoculture is also likely to prove rather effective at preventing world wars.

Across the world, a generation of rising, interdependent youngsters has grown up against a global landscape of no-slack; has learnt to surf the texture of an increasingly manipulative media; and has few realistic hopes for any long-term job security. Most young minds today—minds that increasingly choose not to vote or to politically affiliate—also already recognise that markets and not governments are rightly or wrongly taking control.

As proactive, technology-empowered shapers, the *neXt Generation* are the most valueware perceptive race ever to surf the planet. They won't waste their time, let alone their money, on hardware or software, or on organizational and market structures or processes, that don't add immediate, perceived value. Peace, thin margins, a flexible labour market, and hypercompetition, are all that most young people in the industrialized world have ever known. Yet somehow they have also evolved to thrive on the uncertainty of today's techno-evolutionary free-fall. Indeed as Douglas Rushkoff concludes, our children:

> . . . have already made their move. They are leading us in our evolution past linear thinking, duality, mechanism, hierarchy, metaphor, and God himself toward a dynamic, holistic, animistic, weightless and recapitulated culture. Chaos is their natural environment.[246]

Whatever millennial realities lie ahead, rest assured that our children will arrive there not just before their remaining forebears, but in addition far better prepared. Like it or not, *the kids don't need to get it*, and from that we would be foolish not to learn. More than any other information source, the value aspirations of today's brightest young people reflect the future as we can hope it to be; their minds free ideas yet to be banished down the blinkered and cynical corridors of age, tradition or maturity.

PART III
CONCLUSIONS

8
Millennial
Realities

How we choose to shape our future, and how we choose to realize our dreams, greatly depends upon where we decide to draw the line between present and future fantasy and reality. If we do not believe that some of the predicted developments of the New Age will be achieved, then our closed minds will inevitably condemn them to remain as residents of the Land of Fiction. Across the cyberspace medium, new 'virtual' realities will soon come to be experienced. New social patterns will also emerge, whilst new forms of business organization will rise to craft the information of the world toward the accomplishment of previously undreamt of crusades. In the New Age, Cyber Business developments will not only change how we perceive reality, they will also alter the nature of reality itself.[247]

THE PREVIOUS SEVEN CHAPTERS have highlighted a wide range of technological, social and organizational developments that are going to play a major role in shaping all of our futures. From the Internet to free agent lifestyles; new human–computer interfaces to genetic engineering; Next Big Thing management approaches to on-line lives and the aspirations of Generation X, we have witnessed time and time again how the world of tomorrow is unlikely to prove a simple continuation of today. Embrace them or fear them, incredible millennial realities lie ahead that will radically transform many lives.

The last sentence is typical of those statements made by 'macro futurists' when explaining how we are caught in a whirlwind of startling change. To say that 'millennial realities lie ahead that will radically transform many lives' is almost certainly correct. It may, perhaps, even be of moderate interest. However, it can in no way be interpreted as being of practical *value* for anybody attempting to plan for tomorrow.

Macro futurists, such as Alvin Toffler in his books *Future Shock*[248] and *The Third Wave*,[249] offer sweeping theories of step-change transition to herald the dawn of a whole new era of human civilization. Their analysis is often insightful, sometimes rousing, and on occasions even breathtaking. However, single theories of grand change remain difficult to relate back to the ongoing realities of most day-to-day lives.

To have pointed out to our ancestors of the early 19th century that they were experiencing the 'industrial revolution' would hardly have aided them in planning for the next month or year ahead. Similarly, to contend—with whatever certainty—that we are now entering a 'New Age', 'Third Wave' or 'post-industrial revolution' is of little benefit to those charged with making value-shaping choices today.

Good future gazers need to be effective future shapers. They therefore need to be able to map their visions across a wide range of levels. This is not to say that macro futurists are wasting their time. By getting us to appreciate incredible new possibilities they are most certainly not. However, being able to map grand theory back to at least year-on-year practicalities remains absolutely essential if macro futures work is to be of clear, decision-making value.

Almost certainly because of the academic remoteness of a great deal of macro future gazing, 'micro futurists' are far more plentiful in number (if rarely in name) than their broader-brush cousins. 'Micro futurists' may be categorised as those who dedicate themselves to the cutting- and bleeding-edge study of developments in just one expert field. Micro futurists can usually provide detailed, specific predictions for the future of personal computing, genetic screening, microminiature manufacturing, or whatever their specialism happens to be. However, what micro futurists often markedly lack is a contextual knowledge of how broad developments in many other areas may and will impact on their tightly-bounded corridors of individual expertise.

As a result, many micro futurists fail to appreciate big pictures. They simply don't see the wood or disappearing rainforest for the bark on one of its fascinating trees. Some people may argue that a deep and narrow focus is both inevitable and essential if many expert fields of science and technology are to progress. Up to a point this is clearly the case. For example, few people would want a brain surgeon who was a scientific generalist. However, by creating educational and organizational systems that segment most of human knowledge into deep yet narrow compartments, modern society risks having far too few individuals who ever raise their heads to focus on big pictures.

CAPTURING THE MIDDLE GROUND

In future studies, as in almost every other area of human activity, what is needed in abundance is an effective degree of balance. In a practical context, future gazing needs to adopt a high enough vantage point for it to survey a diverse range of technological, social and organizational developments. However, at the same time, future studies also needs to remain close enough to the ground for it to avoid reporting no more than the emergence of a brave new world. In particular, organizations and other governance structures that focus upon the expert development of micro futures *in isolation* risk major blunders.

As just one example, in the 1980s and early 1990s a great deal of time and money was invested in the development of new *analogue* standards for high definition television (HDTV).[250] Most attention was focused on deciding the best widescreen aspect ratio and image resolution to adopt as a new television standard. Battles were subsequently fought between those parties advocating different systems. Yet unfortunately, by looking only inwards to the worlds of television and cinema, what most developers of HDTV failed to appreciate was that the digital convergence of television and computing would make any single standard for HDTV a non-issue.[251]

Alarmingly, today most developers of digital television are *still* obsessed with agreeing specific image resolutions and aspect ratios. This is in spite of the fact that, as Nicholas Negroponte explains, 'the great gift of the digital world is that you don't have to do this'.[252] Hardware standards for new media are now virtually irrelevant. This is because a few bytes of software code will be able to inform future digital televisions, set-top boxes, and computers, just how to play back any incoming binary data stream.

Other narrow media experts are also caught in similar blinkered-thinking traps. For example, at the time of writing, those preparing for the mass-sale of movies on DVD (digital versatile disk) are still hoping to segment and 'protect' markets geographically. They aim to achieve this by supplying disks and DVD video player hardware using different sound-encoding formats in different territories. Yet any DVD player—let alone any DVD-equipped personal computer—will clearly remain able to read the billions of '1's and '0's written on *any* DVD disk. All that will ever be needed will be some software tweaking to 'break through' the encoding of a 'foreign' disk prepared for a different geographic market-place. It is therefore unfortunate that, as in HDTV development, the television industry—for so long wedded to a mindset of *incompatible* and fixed standards across countries—does not seem to have accepted this basic, technological fact.

The blinkered development of HDTV and DVD video in 'digital ignorance' hopefully demonstrates the wider folly of ignoring macro, 'big picture' future studies. Many other cases abound—with a large number (if not all) being related in some way to the likely impact of the dawning Wired Age. Today, let alone tomorrow, no industry can sensibly develop any sphere of its activities without at least half an eye on the evolving worlds of computing and internetworked digital media. Once any product can be broken down into binary, all of the rules of its old marketplace change.

However, back the other way, cutting-edge digital technologists need to keep a watchful eye on changing social structures and new working patterns if they are to avoid becoming technology rich but value blind. Our emerging, digital global hardware platform must serve humanity and its organizations and not the other way around.

Cutting-edge computer science also needs to keep itself aware of developments in biotechnology and genetics. The scale of the engineering in a great many previously disparate sciences is ever shrinking towards a common, near-atomic scale. As a result, no forward-looking area of scientific or technological development can afford to ignore the potential of nanotechnology, as highlighted back in **chapter 4**.

Across society and the business world, affiliations, industries and traditional boundaries are being decimated almost daily. Often this is as a result of recent micro futures planning that failed to anticipate from where the next wave of competition would come. As also discussed in **chapter 4**, the church is now in competition for our souls with *Star Trek* and George Lucas. Multinational, metabranded corporations are similarly competing with nations for mass human affiliation. Across Western economies, telephone companies are getting into the entertainment business. Supermarkets are facilitating dating. Software barons have their eyes on financial services. And, with the growth of connections to the Internet, everyone and her dog is slowly beginning to see (if not to understand) the potential of cutting out the middleman by electronically niche marketing their product or service directly to a *global* customer base.

Micro futures planning may have worked well in an age in which most markets and technologies were discrete from most others. However, for those companies that survive, it won't sit comfortably—if at all—with the new realities of the third millennium.

In light of the above, this first of my two concluding chapters presents two new conceptual frameworks. These are known as the *Five Facets of Reality* and the *Holistic Lens Model*. The former is intended as an aid in breaking down macro futures transitions. The latter can then be applied to further narrow downward segmented macro transitions to a micro level.

Both models have been piloted for some years as scenario planning tools in a range of business workshops. In particular, both the Five Facets and Holistic Lens models have proved helpful in identifying new business approaches, potential new customers, and new market segments.[253]

THE FIVE FACETS OF REALITY

In line with Alvin Toffler, researchers Michael Piore and Charles Sabel contend that we are experiencing a second industrial revolution, and hence that we are currently crossing a 'second industrial divide'. This signals the ending of an industrial age built upon a common paradigm of mass production. In its place, a post-industrial world is dawning in which more flexible organizations will mass-customise their products and services according to customer demand.[254]

Also highlighting radical transformations from the past to the present and beyond, Japanese Government advisor Shumpei Kumon asserts that human civilization can be traced as having progressed through three phases or 'games'. The first of these was a game of *prestige* in which those with the greatest physical might—the greatest armies and city walls—rose to domination. Next came the Present Age game of *wealth* in which the world has largely been controlled by those with the largest financial muscle, and hence most notably the United States. Today, however, Kumon predicts that humanity is evolving to play a game of *wisdom*. In this, power will accrue to those with the most knowledge, rather than the possession of the greatest industrial scale or fiscal clout.

The implications of the long-standing work of Piore and Sabel, Toffler, Kumon, and a whole host of their contemporaries and followers, is that the environment within which all organizations must operate is caught in a period of bewildering transformation. As already noted, such a proposition is most probably correct. However, due to its breadth it remains difficult to map back to grass-roots business planning. The *Five Facets of Reality* model is therefore an attempt to develop 'downwards' the macro-futures 'past–present–future' grand theory of Toffler, Piore and Sabel, Kumon *et al.* It does this by isolating five continua whose present–future transformations in amalgamation are giving rise to today's radical dawn of the 'post-industrial age' or 'Third Wave'.

By highlighting five discrete if interrelated sets of current transitions, the Five Facets model makes it easier for organizations to understand those most important future changes that may transform their *particular* markets and customer base. In effect, the Five Facets model indicates how five areas of future possibility are being shaped by those key value engines and value perspectives examined in previous chapters. Applied as a conceptual tool rather than as a tight, catch-all framework, the Five Facets model may therefore enable the present-day choice of the most valued and effective realities of tomorrow.

Using a timescale roughly in line with that highlighted by Toffler and his contemporaries, the Five Facets model defines 'the past' as that vast span of human history from the dawn of the early Near Eastern empires right up until the mid- to late 19th century. The 'present' is then taken to comprise the mid- to late 19th century to around the end of the 20th. The 'future' subsequently becomes all of our tomorrows from the early 21st century plus.

REALITY FACET	PAST		PRESENT		FUTURE
Achievement Focus	Awe	▶	Ingenuity	▶	Imagination
Member Status	Serfs/Slaves	▶	Employees	▶	Free Agents
Knowledge Media	Single channel	▶	Multi-channel	▶	Metachannel
Geographic Span	Constrained	▶	Relaxed	▶	Unlimited
Productive Form	Feudal/Craft	▶	Bureaucratic	▶	Organic

Figure 8.1 The Five Facets of Reality Continua

Figure 8.1 illustrates the past–present–future paradigm shift of each of the 'reality facets' of *achievement focus*, *member status*, *knowledge media*, *geographic span* and *productive form*. The following sub-sections also define each reality facet, as well as detailing the nature and involved causalities of each of its two step-change transitions. In particular, attention below is focused on the present-to-future progression of each reality facet, so aiding its later inclusion as an input to the *Holistic Lens* business planning model.

Achievement Focus

The first reality facet of *achievement focus* indicates those kind of accomplishments that any society most values. For example, across the ages, some civilisations have dedicated themselves to constructing great religious monuments. Others have focused on amassing great armies, creating lasting works of art, hoarding riches, or advancing science.

In essence, achievement focus encapsulates the very *nature* of those overwhelming pinnacles of accomplishment that may create value by inspiring widespread pride, reverence, majesty or wonder. As a result, achievement focus also happens to reflect that for which any culture, society, industry or organization is best known, as well as that for which it is most likely to be remembered.

From the past to the present and into the future, the nature of achievement focus has evolved from 'awe' through 'ingenuity' and towards 'imagination'. To explain, the most amazing human achievements of past times were single *physical* wonders of awe-inspiring magnitude. These notably included the Pyramids of Egypt, the Great Wall of China, and Europe's medieval cathedrals. In contrast, since the late 19th century, organizations and society in general have come to value great *acts of cleverness* or ingenuity (and their subsequent consequences) over single physical constructions or epic events.

For example, we remember Thomas Alva Edison not for making one individual filament lamp, but for inventing light bulbs as a practical *concept*. Similarly, the US Moonlandings programme was amazing (and is remembered) as an ultimate act of extreme human cleverness (a footfall on the Moon), rather than for any particular rocket or space vehicle constructed during its nine year duration. In short, across the Present Age, single, tactile wonders of awe have become secondary to an appreciation of the ingenuity of great works or articles of *process*.

However, as Toffler's Third Wave dawns, it is likely that not even mighty works or articles of process—let alone tangible outputs crafted from their application—will be attributed with the most value. Increasingly, technological advancements—in areas from computing and telecommunications to genetics and nanotechnology—are enabling the realisation of almost any dream. In future, it will therefore not be *what* we make (as in the past), nor even *how* we make it (as across the Present Age), but instead *why* we choose to engage in a particular purchase or venture, that will prove most significant and which will hence generate most value.

Already, many 'bad' products and services are just as technically feasible as 'good' ones. Future managers and Future Organizations will therefore more than ever before have to learn to cleave purpose from purposelessness before they commit to any value creation process and subsequent consumer offering. When anything is possible (for a price), so the most prized human and organizational asset is destined to become the creative 'why' of imagination, rather than the mere 'how' of ingenuity, or the 'what' of awe and ancient wonder.

Member Status

The second reality facet of *member status* provides a measure of the type of work role held by the majority. Before mass industrialisation, this was implicitly some form of serfdom or slavery, with workers directly or indirectly owned by their lords and masters. Such individuals therefore had few rights and few (if any) freedoms.

Into the Present Age, as capitalism and industrialisation spread, so the rights of workers slowly increased as the most common member status became that of 'employee'. Freedoms were few—with most workers comprising cogs within great, bureaucratic organizational machines—yet people were paid for their time rather than owned. From the early 20th century onwards, rights and privileges for many employee groups also significantly increased.

However, as highlighted in **chapter 3**, since the mid-1980s a new work/command relationship has begun to emerge based upon 'free agents' who enjoy considerable freedoms, yet who possess few (if any) long-term employment rights. For many 'portfolio workers', temporary employment, zero-hours contracts, part-time working, subcontracting and outsourcing are already the norm. Teleworking is also increasingly common, whilst members of Generation X have never aspired to become long-term, single-company-career employees.

Driven by that harsh, consumer-led economics that now demands high quality on ever-thinner margins, a free-agency workstyle *and mindset* will become a 21st century norm. As Charles Handy argues, no longer are organizations going to 'stockpile people', with the 'employee society' already 'on the wane'.[255] Indeed, in the near future, more and more individuals will both *own themselves and trade themselves* in a labour marketplace that exhibits few rights, if a great deal of freedom to either fly or to fall. Looked at another way, a shift may now be observed in organizational human resourcing away from fixed-employment hours and long-term roles, and towards a far more flexible, task-driven approach for maximizing each labour unit value added.

Knowledge Media

The reality facet of *knowledge media* provides a measure of the means used to replicate ideas over time and distance. It therefore aggregates the synergistic impact and application of humanity's communications media and information storage tools.

Many a traditional wisdom concerning media evolution cites a progression from the spoken word to the written word to electronic audio/visual communication.[256] However, as suggested in the last chapter, in actuality a broader and more subtle progression has probably taken place. Across time, this has involved a past–present–future evolution from 'single channel' media, to 'multi-channel' multimedia, and towards future 'metachannel' metamedia.

Until the mid- to late 19th century, almost all human communications travelled by a single pathway across either time or distance. Usually this involved messages passed by word of mouth, in writing, or in the form of a single picture or other artwork. Then, towards and into the 20th century, printing, photography, radio and cinema all came to be used *in amalgamation* to pass messages between the hours and spaces that separated human minds. For the first time news could be gathered and verified from multiple media sources—from *multi*media—such as newspapers, movie reels, and radio.

By the mid-20th century, the first true multimedia of talking films and television had also become popular as the dominant knowledge media of their age. However, at the dawn of the 21st century, some organizations are starting to become adept at the control and integration of so many multi-channel media in combination that they are finally able to blanket consumers with synthetic, controlled 'metamedia experiences', as opposed to mere multimedia soundbites.

The emergence of a metachannel knowledge media can at first be hard to appreciate. Partially this is because it involves convergent media being applied so ubiquitously that those to whom communications are targeted are likely to be unaware of exactly *where* every component communication is coming from, or at least the extent of the media manipulation to which they are subject. Yet, as discussed in **chapter 3**, already global corporations such as Coca-Cola, McDonalds, Microsoft, Apple, Disney and PepsiCo (to name but a few), have become adept at trading not just in global products, but in exporting global *metachannel experiences* that in turn demand back their homogeneous world brands. As a result, a trade in hyper-real visions down high streets and malls, through the airwaves, across the Internet, upon tee-shirts, via the movie screen, and on product packaging, is increasingly the norm.

As the differences between different market players in all industry sectors continue to diminish, and as most profit margins shrink, so managing metamedia perceptions is becoming of critical importance in maintaining consumer confidence, brand loyalty, and globocultural credibility. What's more, just as the multimedia revolution started one hundred years ago yet did not really take hold before the arrival of television, so potentially the looming ocean of the 'metamedia revolution' will engulf the economies of the world only once true metamedia arrive. These will take the form of highly interactive home entertainment and trading systems, most probably based on low-cost, consumer virtual reality computer links. In the relatively near future, many companies may quite literally be able to part-trade in synthetic experiences over which they will have total control. Indeed, distributed (networked) VR systems will not only be a competitor to the passive multimedia of television. More significantly, they will also upstage the current two-way, interactive media of the telephone and a largely text-based Internet.

Geographic Span

The fourth reality facet of *geographic span* measures the physical area across which any society or organization may achieve influence or domination. In part, geographic span is therefore limited by the third reality facet of available knowledge media. However, technologies and infrastructures for physical trade and transportation also prove a significant constraint on geographic span at any particular point in time.

Many a conventional paradigm for explaining humanity's historical geographic advancement has suggested that regional economies and civilisations expanded outwards due to a developing trade in physical goods. National economies and civilisations are subsequently argued to have further expanded due to increasing political exchanges. Today's emerging global rather than international economy and culture is then popularly associated with a new era of trade in information. Or, as

Malcolm Waters argues, 'material exchanges localize; political exchanges internationalize; and symbolic exchanges globalize'.[257]

The Five Facets model includes a similar evolution from a 'constrained' (regional) to a 'relaxed' (national) to an 'unlimited' (global and beyond) geographic span. However, such an expansion is not linked to a shifting physical–political–informational exchange paradigm. Rather, it is contended that a widening of physical, economic and cultural boundaries has occurred as human societies have been forced to venture farther and farther afield in order to avoid the inevitability of isolated-system stagnation. The Second Law of Thermodynamics (which basically states that all isolated systems will decay unless wider energies are brought into play) therefore becomes the universal determinant of humanity's ever-increasing geographic spread and continued connectivity past, present and future. To cite from *Challenging Reality*:

> . . . human communities from early tribal settlements to entire towns, regions, cities, or even nations, are most sensibly treated as isolated thermodynamic systems. All human communities are therefore inevitably subject to the Second Law of Thermodynamics. As such they will always tend towards decay and stagnation unless fresh physical resources are delivered into their closed boundaries in order to render new and continuing life energies. Consequently, any group of humans who continue to live within a fixed geographic region will face extinction unless they make progress in widening their geographic horizons.

> By constantly moving around, early nomadic communities avoided the curse of living within an isolated environment with a limited resource base. However, as civilizations advanced to construct larger and larger settlements, continual relocation became impractical. Across history, the geographic spans of most successful human collectives have therefore been constantly expanded across lands and oceans in order to commandeer fresh raw material supplies. Just as importantly, patterns of trade between different regions have also arisen in response to the limitations of isolated economic geographies. By opening up trade between two or more isolated communities, each becomes permitted to concentrate upon those productive activities in which it enjoys comparative advantage. The overall effectiveness of resource utilization by many regions in economic amalgamation thereby comes to be improved.[258]

Today, as we look ahead into the third millennium, it is becoming essential to open most human systems as widely as possible. Many markets—for example those in microprocessors or pharmaceuticals—are no longer sustainable on a national or regional level due to the sheer product sales volume needed to cover research and development costs. As highlighted in the last chapter, many human concerns and problems— from global warming to pollution and nuclear proliferation—can also only be sensibly addressed on a planet-wide scale. New multi- and metamedia technologies are additionally empowering individual citizens to share a global technoculture.

The transition from a relaxed geographic span to an 'unlimited' one of globalism is primarily a reflection of humanity's need to expand its present physical and mental boundaries in order to further survive and evolve.[259] A new generation of workers and customers who champion interdependence over independence, and who will probably also soon trade in relationship-rich markets, is also a causal factor in today's decimation of previous geographic segmentations.

Productive Form

The first four Facets of Reality encapsulate historical transitions in humanity's physical, technological and cultural *environment*. The fifth is subsequently left to highlight how organizations past, present and future have evolved, and may continue to adapt, in the face of such change.

Specifically, the progression of the dominant structure or *productive form* of work and its collective organization commences with the 'feudal/craft' based arrangements of pre-industrial days. From the mid-19th century onwards, bureaucratic organizations geared for mass-production then emerged in response to the spread of capitalism and rising Western industrialisation. These huge dinosaurs proved solid beasts upon whose backs to build mighty economies. However, facing the same problems of lumbering inflexibility that wiped out their prehistoric reptilian cousins, bureaucratic organizational giants are today seriously under threat from a new generation of leaner, meaner and more 'organic' productive structures.

Over the past decade in particular, the previously dominant, centrally-coordinated Fordist logic of 'big is best' has come to be seriously challenged by a whole host of more flexible organizational configurations. These often draw upon relatively dispersed resources only as and when necessary, and have variously been labelled as 'dynamic networks',[260] 'internetworked businesses',[261] and 'virtual organizations'.[262]

Regardless of academic label, what most organic productive structures increasingly involve is the effective business application of computer networks and advanced middleware (as detailed in **chapter 2**) to exploit knowledge and creativity across time and distance. Many organic organizations also often globally trade across borderless economies. They additionally tend to flexibly engage a significant proportion of their workforce as free agent individuals. Organic productive forms may consequently be illustrated at the nexus of four interdependent and interacting value creation tools and forces as shown in **figure 8.2**.

FROM THEORY TO PRACTICE

The business value of the Five Facets of Reality model arises from its segmentation of the macro-futures transition of the 'Third Wave' or 'second industrial divide' into five more micro sets of present–future progression. Reading down the 'present' and then 'future' columns of **figure 8.1**, the model highlights how a business environment that championed process ingenuity, long-term employment, multi-channel media, and geographies relaxed only to the national level, sustained largely static, bureaucratic organizational structures. In stark contrast, Future Organizations are likely to have to adopt flexible, organic structures in order to survive in markets in which free agent workers *and customers* most value imagination—the 'why' over the 'what' or the 'how'—and in which trade increasingly takes place in relationship-rich metamedia experiences across an unlimited geographic span.

The last two sentences are clearly in danger of becoming two more of those boasting potentially significant meaning yet little practical value. However, it is important to realize than an understanding of the Five Facets model has already triggered many successful brainstorming sessions for managers from a range of organizations. In particular, with a little thought, most people are able to plot the actualities of their present

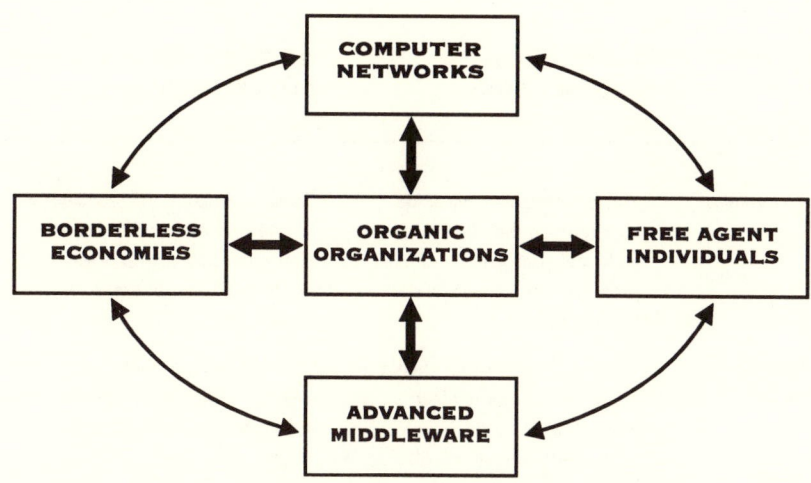

Figure 8.2 Tools & Pressures for Organic Organization

perceptions and strategies against the Five Facets of Reality continua in order to gauge how prepared their organizations may be to create value in the 21st century.

Think of your own organization, as well as choosing two or three other companies of which you have some knowledge. Ideally, select at least two firms from different business sectors. Then, for each company, try to decide whether its *strategies, products and services* most closely align to either the 'present' or 'future' nature of the reality facets of achievement focus, member status, knowledge media, geographic span, and productive form. For your assistance, **table 8.1** indicates how organizations may potentially exemplify a 'present' or 'future' focus with respect to each reality facet.

Even a brief attempt at the above exercise will hopefully focus your thinking towards the *implications* of those millennial realities highlighted by the transitions endemic to the Five Facets model. Trying to classify several companies from different sectors ought also to allow some comparison of the relative strategies that contemporary organizations now exhibit with respect to their vision of, and preparation for, the 21st century.

REALITY FACET	PRESENT-FOCUSED ORGANIZATIONS	FUTURE-FOCUSED ORGANIZATIONS
ACHIEVEMENT FOCUS	Those companies that trade largely upon the cleverness or technical competence of their products, services or productive processes.	Those companies that constantly sell new ideas, or which at least offer older products and services in imaginatively differentiated ways.
MEMBER STATUS	Those companies that expect to continue to employ most workers as long-term employees, and which also expect their customers to exhibit a secure, single-organization career status.	Those companies that have adopted some 'free agent' working arrangements, that understand the new labour realities of Generation X, and which recognise how an increasing proportion of their customers will live free agent lives.
KNOWLEDGE MEDIA	Those companies that trade in products rather than brands, and which do not strongly attempt to synergistically control every form of advertising and other customer or supplier communication.	Those companies that recognise how customers may *affiliate* with a brand that is more than a product, which have blanket media strategies, and which attempt to deliver a purchase *experience*.
GEOGRAPHIC SPAN	Those companies that operate and aspire only locally or nationally in terms of their customer, supplier and/or labour base.	Those companies that market and trade globally, and which hence *either* possess a global scale, *or* which use media like the Internet to electronically disintermediate advertising, sales and distribution.
PRODUCTIVE FORM	Those companies that appear needlessly rigid and bureaucratic in their activities, and/or which 'waste' resources on structures or processes that do not clearly contribute customer value.	Those companies that always adapt rapidly to new market and customer realities, and which have effectively maximised their structure/process organizational valueware overlap.

Table 8.1 Present–Future Organizational Classifications

I believe that those organizations that act and evolve organically with global aspirations, that use metachannel media to part-turn their outputs into experiences, and which offer imagination-first products and services to potentially free-agent customers and workers, will create the most *value* in the future. From your exercise, it is unlikely that you will have identified *any* organization that truly lives and breathes the 'future' classification of all five reality facets. However, from workshop experiences, I would guess that if you have labelled any organization as 'future focused' with respect to most or all reality facets then it is most probably your own. Indeed, workshop outputs suggest that most people *looking out* from within an organization tend to rate it 'more highly' on the Five Facets scales than anybody outside *looking in*. Of course, questioning why this may be can also prove enlightening.

Whether time proves it right or wrong, the Five Facets of Reality model at least presents businesses today with a starting point when attempting to brainstorm present and future strategies in the face of complex technological and social transitions. As a result, it may also be used to generate valuable inputs into what I and some of my colleagues have termed the 'Holistic Lens'.

THE HOLISTIC LENS

The idea behind the Holistic Lens model is that it is pointless to try to plan for the future by applying expert knowledge in only one or a few isolated areas of technological, social or organizational development. In order to segment macro-futures transitions (such as those highlighted in the Five Facets of Reality model) down to a specific organizational level, the Holistic Lens is therefore presented as a four-stage brainstorming 'funnel'. Illustrated in **figure 8.3**, the Lens is intended to be used in a group exercise to focus 'WorldView' and then 'PeopleView' changes down to narrower 'SectorView' and finally specific 'Organization' perspectives.

WorldView changes aggregate all of those broad technological, social and organizational developments that will soon define the business

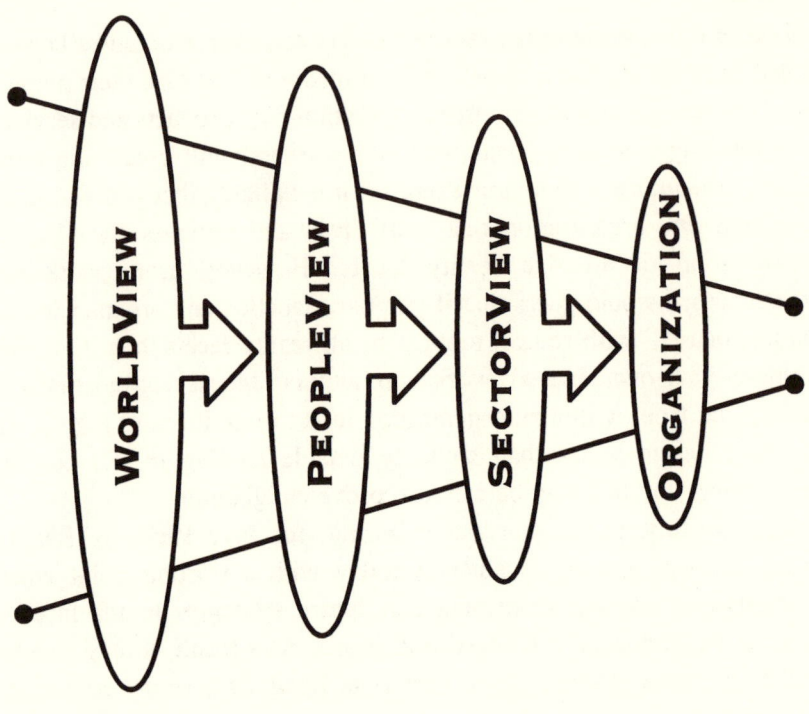

Figure 8.3 The Holistic Lens

landscape against which all companies must trade. PeopleView changes then report the *impact* of looming WorldView changes upon the desires and demands of individuals as citizens within society, and in particular as customers within future markets. SectorView changes next consider what, in broad terms, new customer desires and demands may mean for a chosen business sector. Finally, as the Lens narrows to a micro level, actions, strategies and developments for an individual organization within its particular market sector ought to result.

Key to any successful use of the Holistic Lens is an appreciation that it is the *flows of causality* that may be identified between its 'WorldView', 'PeopleView', 'SectorView' and 'Organization' levels that constitute its most critical outputs. Most managers know their own industries and organizations well. Most can also appreciate the broad thrust of those dawning millennial realities as highlighted, for example, by the Five Facets model. However, linking the two—as I have repeated time and again across this chapter—is for most individuals no easy feat. Yet, when such linkages are made, they often lead to the innovation of radically-competitive new products or services that may greatly enhance an organization's future success.

Table 8.2 provides an example of the Holistic Lens model in action by applying it to the financial services sector. In the far left column, the five broad present–future transitions identified in the Five Facets model are used to 'seed' the lens. Some of the broad implications of these changes for individuals as citizens and customers in the world of tomorrow are then listed under the PeopleView heading. Being inherently generic, these first two columns may initiate the use of the Holistic Lens within a wide variety of business sectors.

Moving across the table, the SectorView column considers what a future world of less materialistic, more autonomous, less passive, more mobile and less secure customers may mean for the financial services sector in general. Finally, listed in the far-right column, are some generic 'outputs' that may lead to new product/service developments and strategies for that particular organization placed at the focus of the lens.

Across **table 8.2**, the linking causalities between different levels of future prediction are hopefully fairly obvious. If asked to develop a business case for any of the three new developments listed in the Organization column, any manager ought to be able to explain how those preceding SectorViews highlight the need for such developments. In turn, changes listed in the SectorView column become 'obvious' once the implications of those entries in the PeopleView column are considered in a sectoral context. The PeopleViews may then themselves be mapped back to an analysis of the broader WorldView changes on the far left of the lens.

In this particular example of the application of the Holistic Lens, the inputs to the WorldView column have also been initially narrowed down via by the Five Facets model. A direct progression from a macro-futures prediction of a New Age or Third Wave, right down to some clear, micro-futures, organization-specific strategies or action points, may therefore be charted in a rational, logical manner. The business value of broad-brush future studies is, incidentally, also thereby demonstrated.

WORLDVIEW	PEOPLEVIEW	SECTORVIEW	ORGANIZATION
TRANSFORMATIONS FROM	*CUSTOMERS BECOMING*	*INDUSTRY IMPLICATIONS*	*DEVELOPMENTS TO CONSIDER*
An *achievement focus* of ingenuity to that of imagination.	Less materialistic (and hence potentially likely to spend more on services).	May imply an increased demand for individually tailored, flexible and tax-neutral products.	New personal pension products including low-cost options to alter contributions and to take contribution holidays.
A *member status* of employees to that of free agents.	More autonomous as individuals.	Also rising demand for financial service products that help facilitate free-agent lifestyles, and which are supported by long-term, one-to-one, two-way-interactive customer-organization relationships.	Mortgage and other loan assessments supportive of self- or short-term employment.
A multichannel *knowledge media* to one based on metamedia metachannels.	Less passive recipients of broadcast media.		
A relaxed (national) *geographic span* to one being more unlimited (global).	More likely to travel, trade and/or live abroad.		Branding of *total financial service packages* ('individually customized welfare states') that integrate a pension, PHI, healthcare, old-age care, etc).
A *productive form* of bureaucracy to one where firms are more organic.	Less likely to have long-term, single-organization ties and long-term job security.		

Table 8.2 A Five Facets Holistic Lens for Financial Services

BRAINSTORMING AHEAD

The application of the Holistic Lens model as illustrated in **table 8.2** may be adapted for a great many organizations and business sectors. However, the model can also be used as a brainstorming tool in isolation from the

Five Facets of Reality. In particular, it can prove useful in isolation as a planning aid or eye-opener in many a business workshop or 'awayday'.

When conducting such an Holistic Lens exercise, the facilitator ought first to ask participants to suggest all of those cutting-edge developments —technological, social or organizational—that they believe may have a significant impact on the way in which individuals live their lives or organizations trade. Such a list usually takes around fifteen minutes or more to generate, may be catalyzed with reference to a book such as *Valueware*, and typically results in a long list of entries. For example, a list of WorldView changes identified under 'technology', 'humanity' and 'organization' headings in a recent workshop brainstorm was as follows:

WorldView technology changes/transitions:

- The Internet
- Digitization
- Nanotechnology
- New battery technologies
- Virtual reality
- Biotechnology
- Software agents
- Space technology
- Global networks
- Unlimited bandwidth
- Superconductors
- Laser technology
- Electronic cash
- Flat screen displays
- Genetic engineering
- Biometric sensing
- Renewable energy resources
- Neural networks
- Miniaturization

WorldView humanity (social) changes/transitions:

- Environmentalism
- World population problem
- Walled communities
- Knowledge of genetic make-up
- Break-up of traditional family
- Increasing Third World power
- Increasing pollution
- Tribalism
- Increasing importance of icons
- McDonaldization
- Expansion of choice
- Spread of drugs
- Cultural globalization
- Increasing life expectancy
- Overclass/underclass divide
- Increased individualism
- Terrorism
- Less leisure time(?)
- Techno (il)literacy
- Global individuals (as icons)
- Global (organized) crime

WorldView organizational changes/transitions:

- Globalization
- Increased competition
- Reduction in hierarchies
- World brands
- Short-termism / capitalism
- Demutulization
- Increased social responsibility
- Increased use of consultants(!)
- Emphasis on environmentalism
- Disintermediation
- Workers seen as individuals
- Flexible working
- More women working
- Organizations becoming more powerful than governments
- More joint ventures & mergers
- Client/server organization
- Outsourcing
- Short-term ownership
- Homeworking
- Deregulation
- Increased litigation
- Pay divergence (Fat Cats!)
- Increasing share ownership
- Increasing customer focus
- Faddism
- Portfolio lives
- Decreasing employee loyalty

Once a list of WorldView changes *in which participants have some ownership*[263] has been generated as above, the next step in applying the Holistic Lens is to ask people to consider how each and all of their identified WorldView changes and transitions may impact upon individuals as citizens of society and as customers of organizations. Again for illustration purposes only, the initial PeopleView list drawn from the above workshop's WorldView outputs was as follows:

PeopleView changes/transitions:

- More choice (access to customised products in global markets)
- Less choice (same brands/products everywhere)
- Confusion / information overload
- Virtual (non-face-to-face) organizational interfaces
- 'Want it now' mentality
- Isolation / alienation
- Anything possible
- Importance of products offering, or packaged with, an experience
- Invasion of privacy in information age
- Innovative new experiences (eg VR)

PeopleView changes/transitions (continued):

- ☐ Lazy consumers (couch potatoes)
- ☐ Demand for low cost and high quality
- ☐ Insecurity and loss of traditional norms
- ☐ Brand dominance
- ☐ Loss of identity
- ☐ 'At my convenience' attitude

Whilst a long list of PeopleView changes may be useful in some contexts, usually in a workshop it is necessary to agree to generically narrow it down somewhat in order for further analysis at the SectorView and Organization level to become practical. For example, following some discussion of the broad PeopleView changes as above, a focused list of future customer changes, transitions (and hence sector/organization challenges) was agreed as follows:

Focused PeopleView changes/transitions:

- ☐ Choice and confusion
- ☐ Insecurity and isolation
- ☐ Virtual (on-line) interfaces
- ☐ Increasing individual consumer determinism
- ☐ Experiences demanded with or as product/service solutions

From commencing the Holistic Lens exercise to arriving at the above focused list took a group of twenty-five middle and senior managers around an hour of sometimes heated argument. From there, smaller groups then developed SectorView and Organization outputs which, for non-disclosure purposes, are not reportable here. However, as by definition the WorldView and PeopleView stages of the Holistic Lens are generic in nature, it is possible to develop the above group's five PeopleView outputs forward in almost any sectoral context. To briefly illustrate how such an exercise may reach a conclusion, I will therefore here apply the five above PeopleView changes to develop SectorView and Organizational outputs for Higher Education.

For the Higher Education sector, the focused PeopleView list almost certainly implies a need to offer clearer yet more flexible pathways through a variety of modular programmes that may be tailored according to individual skill and educational needs. In the future, such programmes are likely to have to include a mix of face-to-face and on-line student contact, in addition to the use of interactive CAL (computer aided learning) multimedia materials. At a broad level, the above list of PeopleView changes also suggests that there will be an increasing need to 'reengineer' the traditional lecture experience in the light of new metamedia developments, and in particular the broadening attention breadth (as opposed to depth) of those more self-deterministic customers of modern generations. Students adept at surfing the information of the world (possibly in virtual reality) may simply no longer sit passively and listen to any single lecturer standing before them for a couple or three hours.

At an organizational level, the above SectorView factors may need to drive the provision of accurate, constantly-updated programme, assessment and curriculum information for students who find themselves interacting with often sprawling and confusing colleges and universities. Within many Higher Education institutions, the development of Internet, intranet and extranet resources, alongside the establishment of curriculum-led on-line virtual communities, may also need to take priority over future expenditure on new physical infrastructures such as buildings.

Higher Education institutions may also need to stop teaching IT *skills* in order that they may focus instead on the use of computer-based digital technologies as a new educational delivery and support *media*. To this end, the fostering of distance learning teaching competencies amongst staff, coupled with a reevaluation of the most appropriate level and use of face-to-face contact required to maintain an effective social/educational environment, is likely to prove critical for those Higher Education establishments that will in future create most value.

In many industries a few organizations may already be identified as exhibiting some of those characteristics that an application of the Holistic Lens is likely to highlight as critical for future success. For example, in Higher Education, the Selye-Toffler University is, in its own words, 'the first World University to link leading professors in three dozen universities to students to offer post-graduate education via the World Wide Web'.[264]

Founded in 1994, Selye-Toffler's Faculty includes around 400 professors spread across eighteen countries. Via this internetworked knowledge resource, the University offers a completely individualized curriculum to every student. As a result, Selye-Toffler claims that every student will spend 'three to four times as many of their total hours on materials and assignments they find relevant and challenging'. They will also never have to 'sit through a one-size-fits-all course (or distance learning lecture) wondering when the interesting parts will occur'.[265]

* * *

MILLENNIAL REALITIES

As think-tank guru Geoff Mulgan nicely explains, 'practice without theory is blind, whereas theory without practice is ignorant'.[266] As a future-gazing resource, the problem with theory commonly lies in its selection of an inappropriate level and degree of complexity. In turn, the problem with using cutting-edge practice to look ahead is that its study can only ever inform us about the recent present. Effective future gazing therefore needs to utilize an amalgam of current theory *and* current practice in order to temper conceptually-rigorous foreviews with some knowledge of recent technologies, organizational activities, and human beliefs.

Whatever some futurists may claim, it will always be possible to dismiss future studies as an activity doomed from its conception because it will 'always be proved wrong'. However, as I argued way back in the Preface, to adopt such a point of view is to largely miss the point. Theories and brainstorming frameworks of future studies—here including the Five Facets of Reality and Holistic Lens models—are primarily intended to trigger people today into raising fundamental questions about tomorrow. By asking such questions, human beings empower themselves to best consciously reengineer what they most desire and value as our collective future. By presenting possible futures from which citizens of today may choose, good models of future studies ought to empower *future gazing as future shaping*.

Any individual or organization that decides (if only through inaction) to sit back and passively accept the future as an inevitable unknown is unlikely ever to glean anything of value from any work of future studies. In contrast, once anybody or any organization proactively chooses to take a hand in crafting their own destiny, they will most probably gain a great deal from theories of tomorrow *and* a knowledge of those cutting-edge practices of today.

The above is not because models of future prediction will magically prove 'correct'; nor because present actualities will somehow reveal valuable nuggets of information. Rather, what proactive rather than passive individuals and organizations already know is that present theory and practice are the only tools available for those wishing to shape their own destinies. The future will remain unknown. However, this does not stop it being crafted from those current theories and practices whose development and study may fire new and powerful human aspirations.

At present, a great many predicted 'millennial realities' remain untarnished by their realization as something real. Unlike the world of today, the landscape of tomorrow can always remain a realm of promise and even magic. As an idea—as a shared conceptual experience—a future shaped by future gazing may therefore help craft itself in the name of many involved expectations.

Whilst frameworks such as those presented in this chapter may trigger important questions, nobody can claim to study in the present those *future* millennial realities that lie ahead. However, this need not stop every one of us today from helping to part-shape with our minds and hands that *potential* world of tomorrow which we may all most value.

9
Hopes & Fears
for Century 21

Empowered by technologies and their fictions of reality, organizations enable us to survive and to realize our passions of collective will. Like the organizations of the past, those of the future will therefore be pillars of successful aspiration. They will also evolve erratically to greatness from competitions of love and hate; from dreams lost and hopes won. As such, Future Organizations will mostly redefine our realities not of economics or technology, but of what it really means to be human.[267]

FOLLOWING A GRUELLING CAR chase and fist-fight in the movie *Raiders of the Lost Ark*, adventurer archaeologist Indiana Jones has his injuries tended by former girlfriend Marion. As she dismisses the battered hero's protests that he can manage by himself, Marion smiles 'You're not the man I knew ten years ago'. To this Indiana weakly retorts, 'It's not the years, honey. It's the milage'.

Time and toil continually weather us all. As a result, no mind can remain constant throughout the duration of a long project. One of the problems in writing a book—let alone a trilogy of books—is therefore maintaining any consistent set of beliefs whilst the work is being undertaken. For authors, books become journeys they may direct, yet whose outcomes they can rarely fully predict. Initial structures can remain solid. However, over time, the passion of writing inevitably rages, twists, falters, and then rages again, if not always in exactly the same direction.

As I open the final chapter of this book, and hence of the Future Trilogy to which it forms the final volume, I mention the above to acknowledge how I have inevitably part-crafted *Valueware* in the dark.

This work commenced with the aim of answering the two questions of 'what is value?' and 'how, in future, will value be created?' In a roundabout way both of these queries have been addressed. Indeed, the last chapter offered two frameworks that drew together my analysis of **Part I** and **Part II**, and which in the process predicted several millennial realities.

However, throughout all of the eight chapters to date, most of this book has focused on the tools and human perspectives likely to *shape* future value creation, rather than on the nature of value itself. Perhaps strangely—and somewhat contrary to my initial expectation—this final conclusions chapter therefore makes a return to the actual definition of value. Further, these last pages also highlight my set of interrelated *hopes and fears for century 21*.

SURVIVAL, CREATION & TOUCH

Whilst preparing this book, I have come to the conclusion that all measures of 'value' stem from the fulfilment of three basic yet co-dependent human needs. As illustrated at the corners of my *Value Triangle* in **figure 9.1**, these are our requirements or desires for survival, for creation, and to touch or be touched. Such needs, I believe, are endemic to the human condition. Their fulfilment makes us 'happy' and keeps us motivated for life ahead. In turn, those collective, economic activities that may assist in individual aspirations to survive, to create, or to touch, are likely to be those that will enable business organizations most successfully to prosper and to profit.

At the apex of the Value Triangle, survival needs span two distinct levels. Firstly, they encompass those needs that lead human beings to value those resources, tools, technologies and social or organizational infrastructures that fulfil our most basic, physiological requirements. At this primary level, we will always value air, water, food, shelter and clothing, as well as the means of their creation, storage and supply.

A level up, secondary survival needs lead people to value those possessions, services and relationships that enable them to *survive more comfortably* beyond basic subsistence. In a strict anthropological sense, it can be argued that telephones, computers, refrigerators, televisions, fax

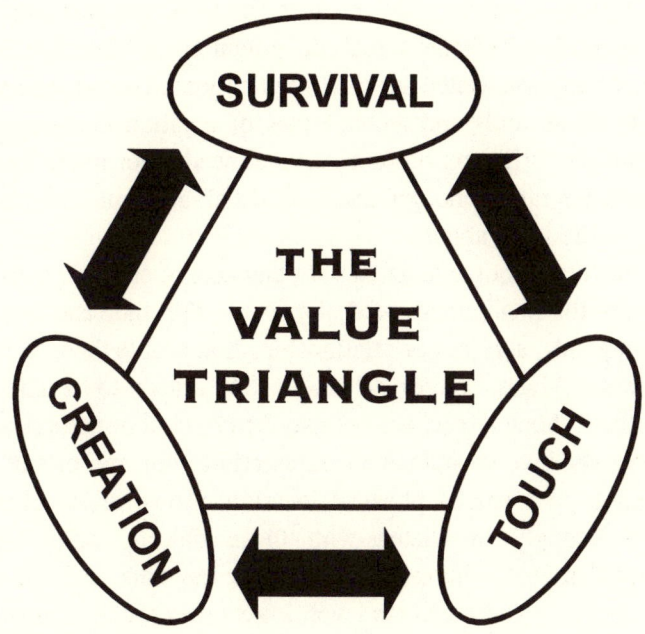

Figure 9.1 The Value Triangle

machines, and indeed a whole host of other goods and services, are hardly survival essentials. However, from a comparative cultural or economic perspective, it also cannot reasonably be denied that most individuals now exhibit multitudinous physical and social needs which, if not fulfilled, will lead to their *survival disadvantage* in the modern world. Beyond being able to drink, eat and shelter in the here and now, it therefore becomes rather difficult to distinguish at a value level between those goods, services and relationships that meet our primary, short-term biological needs, and those which in turn enable our 'survival' to a greater degree of social, cultural or economic comfort.

The second corner of the Value Triangle highlights our need to engage in acts of creation. It therefore indicates how value is commonly attributed to all of those goods, services and relationships that enable creative acts and outputs to occur. On many occasions, individuals need to create in order to survive. However, for most people, there still remain creative aspirations whose actual or potential fulfilment are valued regardless of any associated survival implications. Indeed, many people continue to value tools and technologies of creation that may actually detract from the fulfilment of their survival needs. For many individuals survival alone is rarely enough, and indeed risk-all gambles are for some people what life is all about.

When survival is not associated with any acts of creation, it is usually supportive of the final human need of 'touch'. This indicates how people value those goods, services or relationships that enable them to reach out and influence others, and/or which enable others to reach out and influence them. Such a reaching out may be physical or mental, and may most obviously involve immediate interaction with one or many other human beings in a shared physical location. However, touch value is often also strongly associated with those objects, relationships or experiences that enable human reciprocity across time or distance.

For example, at an immediate level, lovers may value a caress, whilst a sculptor may value those tools, and that time and knowledge, that enable her to craft a work of art through which she may touch other minds. In turn, parted lovers may value past memories of each other, and in particular objects that may trigger such memories. In a similar fashion, the future owner of a statue is likely to value the way in which its physical presence allows the distant inspiration of its sculptor to reach out to her across time and space.

As indicated by the two-way arrows in **figure 9.1**, each corner of the Value Triangle can assist in supporting the two others. It is generally only possible for an individual to touch another (or to be touched back), if they continue to survive.[268] Touching one or more other human beings is also often only possible through some act or acts of creation. Similarly, most acts of creation, or those tools, resources and relationships that empower them, tend to rely on both survival and touch. In turn, it may be argued that we only value having our survival and survival comfort needs met so that we may engage in acts and experiences of creation and touch.

With not too much thought, most people can rapidly attribute at least one of the Value Triangle's three value measures to most of those items, activities or relationships that they personally perceive to be of value. Initially, usually just one value measure tends to spring to mind. However, after a few seconds of thought, a second frequently also enters the head.

For example, when I first conceived the Value Triangle, I was looking up at an old lamp shade. My initial assessment was that I valued it solely as an object that increased my comfort survival level by hiding the glare of a bare bulb. However, I quickly realised that I also value the shade as a possession that has hung in various rooms in most of the residences in which I have lived. In addition to its 'survival in more comfort' immediate value attribute, the shade hence carries with it a sentimental 'touch' value component, as it may trigger personal recollections of other times, people and places.

What the above example hopefully reminds us is how and why different things may be valued differently (or not at all) by different people. Recognising how different individuals—different citizens and customers—attribute value due to different Value Triangle associations will be of an increasing importance for those Future Organizations attempting to create most value. To further assist in an appreciation of the Triangle model, **table 9.1** provides a few more examples of how survival, creation and touch need fulfilment may be attributed to some items, activities and relationships of commonly perceived high value.

VALUEWARE AHEAD

What I wish to propose with the Value Triangle is that any object, service, person, relationship or experience *that is valued* has to be at least part-fulfilling a basic human need to survive, to create, or to touch or be touched. Further, I would suggest that those objects, services, persons, relationships or experiences attributed with most value will be those that help to part-fulfil two or even all three of our needs for survival, creation and touch in amalgamation. Organizations seeking themselves to survive, to create, and to interact with or 'touch' their own environment, therefore need to address how they ought to act in order to fulfil one, two or three of the human needs highlighted at the corners of the Value Triangle.

ITEM, ACTIVITY OR RELATIONSHIP	SURVIVAL, CREATION OR TOUCH VALUE COMPONENTS
Television	Increases **survival** comfort due to the **touch** of other people and their ideas across time and distance.
Personal computer	May assist a user to **create** in a variety of media, to **touch and be touched** by other people and ideas through network communications, and/or to **survive** (more comfortably) if used to organize one's life and/or to engage in business activities.
Evening out at a restaurant, bar or club	Aids physiological **survival** due to the intake of food and liquids, as well as enabling one to **touch and be touched** by fellow human beings in a process that may also increase one's social/cultural comfort **survival** level.
Attendance at a sports event	Enables one to be **touched** by the activities of others—both participants and fellow spectators—and usually to **touch** others as one becomes part of a cohesive or adversarial social climate that increases one's **survival** comfort.
Sex	An act of **touch** that can also enable race **survival** through a process of biological **creation**.
Love	The **creation** of a mutual **touch** affiliation that enhances and empowers many aspects of **survival**.
Money	A medium of exchange that, through a common desire for an ordered system of **survival**, enables the barter, control, alteration or acquisition of physical and sometimes emotional resources that may result in **survival**, **creation** and **touch**.

Table 9.1 Objects, Activities & Relationships of Common Value

In light of these propositions, general 'answers' to this book's two broad questions of 'what is value?', and 'how, in future, will value be created?' may begin to be crafted. Firstly, in answer to the question of

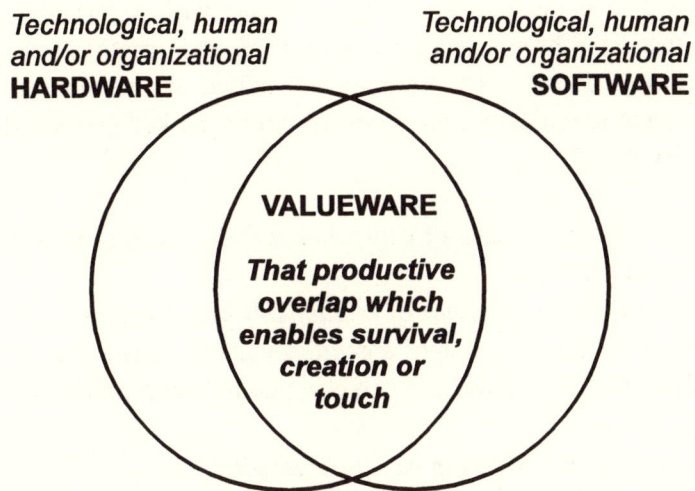

Figure 9.2 Generic Valueware Defined

'what is value?', I have hopefully now validated a reply that value is *that which enables survival, creation or touch*. Secondly, by combining this definition with **chapter 1**'s isolation of 'valueware' as *that productive overlap of technological, human or organizational hardware and software that may actually do something useful*, a generic statement concerning future value creation may also be made.

Specifically, it can reasonably be claimed that all sources of future value creation will be attributable to *overlaps of technological, human or organizational hardware and software that enable survival, creation, touch, or some combination thereof.* All of these definitions and conceptual combinations may subsequently be encapsulated in a single, generic valueware diagram as illustrated in **figure 9.2**.

Those concepts and definitions encapsulated in **figure 9.2** may now be used as tools in assessing the 'value' of all conceivable present and future value engines and value perspectives. Indeed, they will be applied below to enable us to reflect more carefully on some of our previously highlighted means of future value creation, such as computer networks, new middleware software, and virtual communities.

As these final reflections take place, I also wish to recall five key themes that have been interwoven across the eight previous chapters. These, I suggest, all deserve significant individual and organizational attention, and specifically encompass the following key hopes and fears for century 21:

☐ The fear that individuals or organizations will become *technology rich but value blind*.
☐ The hope for the emergence of *relationship-rich markets*.
☐ The hope for the evolution of *a gentler mode of capitalism*.
☐ The hope for a global mindset that champions *interdependence over independence*.
☐ The hope for *future gazing as future shaping*.

By examining this set of one fear and four hopes, the following sections highlight how those value engines and value perspectives explored in previous chapters may in future impact on individual desires to survive, to create, and to touch or be touched. Further, as the basis of a proactive mindset for future shapers, they may also suggest how organizations can be transformed to play their most effective role in meeting widespread human needs.

TECHNOLOGY RICH BUT VALUE BLIND?

Last summer one of my MBA students wrote a dissertation that explored problems with new information technology systems in Brazil. In one of our meetings he told the tale of a large bank that had recently invested in its first automatic teller machine (ATM) or 'cashpoint' network. At a cost of many millions of dollars, this had unfortunately reaped few of its hoped-for benefits.

The problem with the new ATM system was not that the involved technology did not work as planned. Rather, what the bank had somehow ignored when developing its new system was that a high percentage of its customers were illiterate. Once the ATMs were installed, counter staff therefore had to spend large amounts of time helping customers to use them. This in turn negated one of the main purposes of the new system, which had been intended to reduce the number of human tellers required to run a branch.

This story provides an horrific example of an organization that became *technology rich but value blind*. Whilst alarming, the tale is also hardly an isolated one. Indeed, as I bemoaned when I introduced the concept of valueware back in **chapter 1**, today a very large number of individuals and organizations appear to invest recklessly in information technology systems that deliver little, if any, real value.

As we have seen across this book, many of the incredible technologies of the Wired Age may allow us to create value in new ways. Unfortunately, the availability of such technologies also risks turning many individuals and companies into hardware and software junkies capable of squandering more and more of themselves and their profits away.

Virtual working practices like telecommuting, virtual organizations with minimal physical infrastructures, and on-line virtual communities, are now all becoming commonplace. They have also only become *possible* in recent years due to the development of ubiquitous computer networks and middleware software tools. It is therefore perhaps hardly surprising that many companies are now eager to join that rapid, organizational dash to exploit such technologies to their maximum. However, what such firms ought not to forget is that, to date at least, almost all business initiatives solely involving the cutting-edge application of new, on-line technologies have failed to deliver even the *hope* of a cost-covering reward. Not least this has proved to be the case for almost all Internet pioneers. Even many of the service providers that link people into the Internet's global network have yet to make a profit. Don Tapscott's wry comment in *The Digital Economy* that there remain 'more prophets than profits on the Net' therefore ought not to be ignored.[269]

The dividing line between effective, value-creating new technology adoption, and value-negative new technology overkill, will remain both blurred and in motion for years if not decades to come. Almost certainly, in the first half of the 21st century, a great deal of technological hardware and software will therefore continue to be wasted outside of value-positive valueware overlaps. Successful Future Organizations will subsequently be those that learn to monitor their surplus technology investments more successfully than their competitors in order to waste the least. One way they may do this is by constantly questioning how each combination of technological hardware and software they employ may be enabling their customers or workers to more comfortably survive, to create, or to touch and be touched.

In undertaking such a questioning, we would all be wise to temper any evangelism for new on-line technologies with the knowledge that too many hopeful soundbites have already been too greatly believed. As thought tools and drivers of action, soundbites such as 'technology can take the place out of workplace' can prove of significant, proactive value. However, it also needs to be appreciated that the ability to negate distance with the technologies of cyberspace will remain a little more limited than many technoholics and telecommunications companies may ever choose to admit.

For a start, it is becoming increasingly clear that many individuals remain economically and socially bounded by the linguistic peculiarities, traditions, and culture of a non-transcendable local geographic location. Today, anybody with an Internet connection can join any public access virtual community on the planet. However, as many non-US citizens quickly discover, such on-line meeting spaces remain largely *middle-class American* rather than *global*, exhibiting as they do many of the associated traditions and language peculiarities of a hyperreal American Way.

With the above observation I hope not to imply that to have how one lives, thinks or talks 'constrained' by local geocultural boundaries is in any way wrong. Yet, as a British participant in several US-based virtual communities, I never fail to be struck by the linguistic and cultural rigidity of such communities' supposedly 'transparent' geographic bounds. Words, education systems, and even public holidays, tie us to specific geographic locations as surely as our physical bodies. They hence create certain barriers between people who live or were raised in different regions of the planet. Granted, the tools of the Information Superhighway (for many years labelled by the Americans as the *National* Information Infrastructure), enable such barriers to be mostly overcome. However, what I wish to highlight here is that the 'freedom' from geography so often associated with networked technologies is on occasions only a 'total' freedom for those who share roots and a background in a common geographic, cultural and linguistic location.

When questioning the value-creating potential of many on-line developments, it also needs to be remembered that significant psychological distances have yet to be bridged by technology alone. Whatever the champions of teleworking, virtual teams or virtual communities may preach, distance still introduces tension into any relationship lived or worked apart. Contact by phone, e-mail, video-link, and perhaps even VR-link, is likely to remain less frequent, more regimented, and probably less multisensory, than any other form of exchange involving two or more co-located human beings.

All of those communications we undertake with other human beings at a distance require our brains to work overtime at filling in the gaps. Such spaces in those experiences we yearn to share are created by the deficiencies inherent in all present and likely future communications technologies, as well as by those periods of time we have lived apart. There are no chance meetings in cyberspace. Even family members or friends only pick up the phone at an habitual time or when one party or the other desires a communication. And this makes communication at a distance very different from that which continually happens by chance proximity in almost all homes and offices. Relating to anybody at a distance will therefore never prove as *comfortable* an experience as meeting with them in the 'meatspace' of the face-to-face. Computer networks will hence remain value-limited in fulfilling some of our most basic human needs of survival comfort and reciprocal touch.

Due to the tensions inherent in remote communication, all successful virtual communities are likely to require the 'augmentation' of on-line member relationships with 'real' meetings in the flesh. Similarly, most on-line business activities will always be able to be augmented 'in real life' or 'IRL'.

It used to be said that absence makes the heart grow fonder. This may, perhaps, remain the case. However, what we can today report with more certainty is that geographic distance makes any relationship more stressful. Value creation opportunities are therefore likely to become increasingly apparent for those organizations prepared to strip technology away. As demonstrated so powerfully by Generation X, the Wired Age is more than anything about connecting people to people. The most value-

sighted organizations will therefore be those that recognise how even the 'best' technology will always have significant social limitations.

RELATIONSHIP-RICH MARKETS?

Moving from the negative to the positive potentials for value creation in the years ahead, my first hope for century 21 is that we are currently witnessing the emergence of *relationship rich markets*. A large proportion of industrialised economies today are awash with hyper-disposable goods and an over-abundant supply of services. It is therefore a shame that in parallel more and more of their citizens live alone, watch alone, consume alone, and die in isolation. Increasingly, what many individuals are lacking in their lives is the ability or even potential to touch and be touched. My hope for relationship-rich markets is therefore for economic systems capable of selling people back to each other.

Business organizations across the 20th century have become rather good at helping their customers to fulfil their survival needs in greater and greater material comfort. Tools and media that enable people to create are also in an increasingly abundant supply. Indeed, more people now engage in DIY, gardening, video making, and a whole host of other creative activities, than ever before. However, what modern markets remain rather poor at is meeting the aforementioned, burgeoning needs of many individuals to touch and be touched.

To nowhere near the degree of even twenty years ago are employers, nations, local communities, religions or extended families helping to fulfil the touch needs of many lives. Clear opportunities therefore exist—and will probably continue to grow—for business organizations to augment their goods and services with a social element. Indeed, many companies may increasingly be able to package affiliation as a product in itself.

As discussed in several previous chapters, some organizations are now trying to build virtual communities over the Internet. Their hope is that they will be able to use social ties to foster long-term consumer loyalties to their on-line information and service content. In this pursuit some companies will almost certainly succeed. This should in no way be taken to imply that *all* tradable human touch needs will be able to be met on-line. As noted in the previous section, relationships fostered electronically over physical distances tend to be more stressful and less fulfilling than those mediated in realspace alone. However, the limitations of cyberspace interaction will far from totally negate many emergent opportunities to generate touch value for individuals via on-line media.

A part-social trade in human or organizational affiliation may become a necessary prerequisite for many on-line business that hope to foster a loyal and long-term customer base. Today, other customers can transform a bar from a place where shelter and refreshment 'survival' needs are met, into a 'touch retreat' where they can socially interact with others. In a similar fashion, one-to-one and one-to-many electronic interactions on-line have the potential to value-augment Internet shopping malls and a whole host of other cyberspace business arenas. Indeed, even the text-based virtual communities of the late 20th century have transformed the web browsers of many individuals from *information access tools* into *habitual social places*.

Of course, not all organizations aspiring to meet the increasingly unfilled touch needs of a free agent, affiliation-starved population will need to venture into cyberspace. Granted, usually the greater the number of communication channels available between consumer and organization, the more successful any 'affiliation product' setting out to meet human touch needs is likely to be. However, the degree to which individuals can buy-in to a psychological home is likely to prove far more critical in most relationship-rich marketplaces.

As previously noted in **chapter 3**, to date very few organizations have succeeded in building strong affinity products into which consumers may achieve a high degree of buy-in. Those that have, including Lucasfilm with *Star Wars* and Paramount with *Star Trek*, have often been in the film or television industry. Yet this does not have to remain the case.

Already boundaries between goods, services and relationship/affinity product industries are beginning to blur. Pepsi's $2bn deal with Lucasfilm for the rights to use *Star Wars* to promote its beverages provides just one example of a large-scale, physical-product/relationship-affiliation tie-in that may point towards many more relationship-rich economic potentials ahead.[270]

The mass global affiliation that Paramount has managed to sustain long-term to *Star Trek* is another example of a relationship-rich product that ought not to be reasonably ignored. Tom Mazza, an Executive Vice President at Paramount, describes 'The Franchise' as having a 'very theological umbrella hanging over it . . . [that] . . . asks viewers to buy into a lot of belief systems'.[271] Similarly, Brannon Braga, the Supervising Producer of *Star Trek: Voyager* argues that the multi-billion-dollar, 'collective consciousness' *Star Trek* universe:

> . . . is mythos. It represents something positive that people need. It has persevered long enough that parents will pass down an interest in *Star Trek* to their children. The characters and the morality tales are so important and appealing and hopeful. We need mythologies. We need heros to come back to again and again. In the American culture we don't have that. Except for *Star Trek*.[272]

Having cited the above, I would hardly like to suggest that the greatest market opportunities ahead will all involve fulfilling the touch needs of the human psyche to boldly go. However, many organizations that have become so expert at selling *single* goods and services in *single* instances may still have a great deal to learn from some of those metamedia empires of science fiction entertainment. After all, Paramount, Lucasfilm *et al* have successfully found a formula that not only fulfils a powerful human need to affiliate, but which in addition sustains a highly dedicated loyalty to a long-term brand.

A GENTLER MODE OF CAPITALISM?

As the organizations that consumers purchase from become more important in affinity terms than many of those they work for, so my second hope for century 21 is for the emergence of a *gentler mode of capitalism*. As also initially discussed in **chapter 3**, it may well be that it is not *how* we trade, but rather *what* we trade, that is in need of review if collective humanity is to reengineer a brighter, more sustainable future. Indeed, as and if we evolve from a trade in physical 'survival' and 'creation' goods and services, and towards markets richer in traded relationships or 'touch', so the environmental cost of mass capitalism can only decline.

In the May of 1998, a study from the Worldwatch Institute reported the hardly surprising fact that the world was economically richer yet environmentally poorer than a year before. As Worldwatch president Lester Brown noted, whilst the global economy had expanded in 1997 at a near record four per cent, the same year had also been one 'of disturbing new signs of environmental stress'. Rainforests that had burnt for months had irreversibly damaged rich ecosystems. Ice caps were melting, whilst global temperatures, CO_2 concentrations, and carbon emissions, had climbed to record highs. During the course of 1997, the human population of the planet had also risen by over eighty million, with well over three hundred cities now boasting more than a million citizens.[273]

Nobody can seriously believe that, even in the medium-term, a continued, manufactured-goods-driven global economic growth can prove sustainable within our closed planetary system. Hard choices therefore need to be made now if future basic survival needs are to continue to be met for even the currently-fortunate rich minority. Worldwatch's 1998 study did more positively report a growing investment in wind-driven power generation. A rising trend for taxing environmentally destructive activities, rather than income, across a handful of European nations was also noted. However, there can be no doubt that these and similar planet-friendly initiatives alone will do little or nothing to break our current cycle of increased physical consumption and accelerating resource depletion.

Quite simply, the hard fact of capitalism is that it is self-fuelling. Demand feeds more demand. That's how the system works. Therefore, as long as our markets are geared towards selling physical goods and physically-manipulative services, the more and more resources the human race will continue to consume and to waste. The only 'way out'—beyond opening our horizons to the vastness of space[274]—therefore has to involve a fundamental shift in that *type* of consumption which we permit to continue to self-feed.

New relationship-rich markets that economically substitute group affinities and traded human interactions for physical products or services have the potential to prove a part of our salvation. Indeed, one may reasonably argue that those extended groups and structures—such as local communities, extended families, nations and long-term organizations—that used to meet so many touch needs have only eroded *because* 'me-first' market forces have been allowed to accelerate resource-intensive survival and creative individual wants to the fore.

To get back what many now realise we have lost or are losing, we probably need to place a price tag on many of those identity and affiliation social comfort blankets that we used to think of as 'free'. Like it or not, we have created a society and culture—and not just an *economy*—that measures and allocates almost everything of human value in money terms alone. To heighten a broader fulfilment of our touch needs, we therefore may have no choice but to reflect our affiliation wants as economic wants in the global marketplace. It also need be no bad thing that 'the best things in life' may all end up priced in dollars.

We simply need to keep remembering that capitalism is self-fuelling. The system thrives on demand fuelling more demand. The prize for turning human affinities into dollar-mediated, mainstream relationship products will therefore be an increase in their supply and a fall in their price. In parallel, any move to tweak our economic system to value touch, in addition to survival and creation, *has* to lead to a decrease in resource-draining global consumption.

To advocate placing a monetary value on that social something which actually marks us as 'human' may appear both heartless and cold. Yet, by learning to trade in those connections we collectively so desire amongst ourselves, we may double-win by improving our social quality of life *in addition* to decelerating our rate of resource depletion.

INTERDEPENDENCE OVER INDEPENDENCE?

The strongest indication yet that the above hope may prove more than a pipedream can be found stemming from almost any study of Generation X. Today, there finally appear to walk amongst us a generation who value *interdependence over independence*. My third hope for century 21 is therefore that this most powerful value perspective will spread. As and if it does, so most individuals and organizations may finally realize how their individual ability to survive, to create, and to touch or be touched, has to partly depend upon their neighbours.

Humanity has become a hive organism. Indeed, my intention in championing the hope of *interdependence over independence* is to highlight how, whilst mass connectivity may support mass global individualism, it also powerfully serves to illustrate the interwoven fragility of the global creature we have all become. As Gregory Stock argues in defining the single, collective human and technological superorganism of 'Metaman', all individual human beings are today also interdependent 'cells' within a wider, cybernetic whole. As the 21st century dawns, increasingly no individual, no organization, and no nation, can survive, thrive, create or touch in isolation. As Stock concludes his study of our collective, living machine:

> Metaman affirms that we are all connected—giving to and drawing from one another as we participate in a momentous step in the evolution of life. Stone tools came from Africa, writing from the Tigris–Euphrates valley, paper currency from China, the steam engine from Britain—all humankind has played a part in bringing Metaman into being. Together we can exult in this shared accomplishment, try to solve the immediate problems at hand, and look with anticipation to the amazing future stretching before us.[275]

Whilst the above may appear fanciful and even glibly idealistic, traditionalists ought not to ignore the way in which network connectivity and globocultural convergence are bringing more and more of humanity together. Through shared music and television, computer games, global news, and the Internet, those young people now entering the world of business organizations harbour a fresh and collective value creation mentality. Many appreciate how their actions will both influence and depend on the actions and reactions of others as never before. Indeed, as discussed in **chapter 5**, some stakeholder-led initiatives in value based management already partially reflect this fact.

Around the globe, enlightened boardrooms—as well as information surfing youngsters—are awakening to a world of inescapable cultural and economic cause and effect. Increasingly, the freedom to be an individual—be it an individual person, organization or nation—will have to be tempered with the responsibility inherent upon any entity whose actions part-determine the survival and comfort of a wider whole.

Whilst primarily highlighting the symbiotic connectivity of billions of human beings today, any championing of interdependence over independence may also remind us that we remain irrevocably linked to others across time as well as distance. The actions and achievements of our ancestors define to some extent who we are and what we may become. Our ancestors seeded our culture. We may therefore never be 'free' of their influence, failures and accomplishments. Indeed, as we grow older, we may all individually realise how advancing maturity involves learning to live in one's own shadow. As we ourselves, our civilizations, and our organizations age, we must therefore face the sometimes painful fact that our futures will be defined not just by possibilities ahead, but by certainties past.

Just as history and culture link individuals and their creations back into time, so our personal and collective actions today may also weigh heavily on those to come. Across the 20th century, future generations have often been neglected stakeholders in our planning. In championing interdependence over independence, we should be mindful not to forget this fact. A life of truly fulfilled survival, creation and touch needs has to involve more than personal satisfaction in the here and now. Or as I suggested in *Challenging Reality*, one key mindset element for the Future Organization may be *to live for today, whilst planning for tomorrow, and building for the future.*[276]

FUTURE GAZING AS FUTURE SHAPING?

When introduced to an audience as a futurist, I have become used to one of two typical reactions. The first is the wry demand to predict next week's lottery numbers. The second, usually voiced with an even more cynical dismissal, is that all future gazing and future studies are a waste of time. After all, as so many critics continue to argue, nobody can predict the future. 'Absolutely not', I usually reply. 'But that doesn't stop us from predicting a whole range of possible *futures* from which we all may choose'.

My final hope for century 21 is for *future gazing as future shaping*. With this I intend to infer that all *valuable* works of future studies must have some proactive potential. Like many other groups of academics, a great many futurists appear to write and preach solely to each other —effectively to the 'converted'—rather than spreading their wings within other more cynical but also more critical spheres. Such inward looking futures work is, I believe, largely a waste of time. It may fill bookshelves and academic journals. However, it almost always fails to make a difference.

As a futurist—a label, I might add, I never went out to seek—I have no wish to endlessly *describe* how our lives and economies might turn out in years to come. Rather, from what I hope is an educated position, I wish to inform others about what I believe are the most value-laden *options* for times ahead. In the light of such future scenarios, present day individuals and organizations may then start to make *choices* that will directly and positively shape all of our years and decades ahead.

Proactive futurists—futurists who seek future gazing as future shaping—ought to present themselves as technological and organizational catalysts and collage artists. Future studies is most usefully about weaving webs of possibility, raising questions, and crafting mental models which may positively impact on the actual *behaviours* of key future shapers such as managers, research scientists, and politicians.

Visualising possible futures is an art in itself, if a complex and mentally abrasive one that is hence either undertaken narrowly or not at all by most planners. It is easier to believe that only things in one's own area of expertise are about to change, rather than to accept (as discussed and modelled in the last chapter) that there may be far broader transitions afoot. Nobody can predict the future. But many people with free-thinking creative visions may shape it. And some actually will.

There are—most probably—only two certainties in life. The first is death, and the second the fact that until death occurs we are all going to spend the rest of our lives in the future. As a consequence, we all have an incredible incentive to turn future gazing into future shaping.

So does a potter 'predict the future' when she weighs up a piece of clay? Or an architect gaze foolhardily ahead when drafting building schematics? No. Rather, what they both attempt is to *future gaze* in order to help themselves or others to *future shape*. Nobody can 'predict' the future of a lump of clay. Yet a potter can plan its many possible craftings on the wheel. With such 'foreseen options' she may then act with her skill to turn the clay into a bowl or a plate or a mug. Similarly, an architect may *future gaze* by drawing up a range of plans that he and others may shape into the actuality of a future building. In a generic nutshell, future gazing becomes future shaping when and if it brings the possibilities of the mind's eye into sharp reality.

As Dr Michael Moynagh, an advisor to The Tomorrow Project, informed me when we first met, 'we have to find the psychic energy to start thinking about the future or we are all in trouble'. And today this need not prove as difficult as it sounds. With those new, technological and social tools of cyberspace, we have in our heads and our hands the means with which to future gaze and to future shape our collective tomorrows as never before. By learning to recognise *future gazing as future shaping*—whether it be in planning a new shelf in the kitchen, a new form of organization, or a whole new economy of relationship-rich markets—we may all play a role in defining our most *valued* way ahead.

* * *

ENDINGS & BEGINNINGS

There comes a time when any well-chewed research study, mental spark, or even vague reflection, has to be cast forward into the messy uncertainty of reality if it is ever to be judged a failure or a success. Ideas—be they the product of corporate research and development, boardroom brainstorms, academic debate, or Sunday afternoon daydreaming—can do nothing but die if they are left forever in the dark sanctum of the human mind.

The economy of the future will probably above all else be an *ideas economy*. However, this will not mean that raw ideas in isolation will ever have much value.[277] Nor, in a 21st century in which we seem destined to increasingly value imagination over ingenuity or awe, are those products, services or even relationships resulting from good ideas likely to be attributed with significant worth. Rather, as is already becoming apparent in so many markets, what will be most valued in our emerging ideas economy will be those technologies, those individuals, those organizations—those *valueware overlaps*—that will *transform ideas from mindspace to realspace or cyberspace and beyond.*

The economy of the future will be an economy that will more than anything champion action over inaction, and that will more than ever before reward the speed, flexibility and commitment of those daring to surf. To survive, to create, and to touch and be touched, future individuals and organizations will have to get used to taking constant initiatives. They will also have to continually internetwork and to promote themselves in somewhat abrasive marketplaces awash with dynamic global cultures. No longer, as in rose-tinted days gone by, will one or a few good ideas support any individual, organization or industry for life.

Find an analogue clock or watch with a second hand, or else a digital timepiece with a seconds display. Now watch the 360° rotation of the second hand, or the advance of the digital digits, for a complete minute. Go on. For once, when encouraged by an author to put down a book and do something, go ahead and actually do it. There. Wasn't that a complete waste of time? Yet watching minutes, hours, days, months or even years count by is the game that many individuals and most organizations continue to play. In a world in which the rate of change has barely got out of first gear, procrastination continues to rule OK.

In a new ideas economy of relationship-rich markets and mass global interdependence, a mode of capitalism may soon emerge that proves gentler on our planet and even ourselves. However, even when and if it does, there can be little doubt that the harsh whip of time will continue to lash across us more sharply and rapidly than ever before. Often an idea today will have to be transformed from mindspace to realspace or cyberspace tomorrow to have any hope of creating any value. Perishability will become a characteristic of a growing number of products, services and traded relationships. In any race to create value, clock watching will be out. Period.

In our dawning future, individuals are increasingly likely to place a greater and greater value on 'free', unpressured time. Time to take care of their physiological and psychological survival. Time to create. And, perhaps most importantly, time to touch and be touched. As the words of Douglas Rushkoff reminded us so clearly in **chapter 6**, 'people are coming to value time itself—time for contemplation, and time for non-intermediated contact with other human beings'.

As we enter the third millennium, today's accelerating convergence of technology, humanity and organization looks certain to continue. Hidden from view before us there now lie the first thousand years of human history that may be dominated more by markets and capitalism than by the politics, feudalism and warring religions of so many earlier centuries. We may already clearly see how joint social and economic technologies, such as those of the Internet, may both simultaneously enrich and degrade our quality of life. Such parallel enrichment and degrading can be noted to have accompanied the growth of market economies and capitalism since the Industrial Revolution. A parallel giving and taking away is also likely to continue as developments accrue in metamedia, virtual reality, genetic engineering, and new, free-agent lifestyles.

If there is one historical lesson that we may learn about value, it is that whenever the new has augmented the old with additional value, then somewhere else it has also taken value away. Blinded by fresh magic, we often don't see any value loss at first, and sometimes not until it is too late. Yet this timeless historical fact ought not to make us fear any future technological, social or organizational development.

Adding and taking from ourselves is what the dynamism of evolution that keeps us alive is all about. Humanity thrives on change. Going back has always been creative death. The loss or ending of anything worthwhile usually also provides us with the exhilarating if frightening hope of a new beginning.

Epilogue
Technology Rich, Value Blind?

THEY LAY ON THEIR backs under a solid, blue sky. There were no clouds, birds or even suns in this world. They had decided to soak themselves in all of the processing power available.

The grass—the ground—was an infinite, flat plane of monotone green with just the vaguest hint of a fractal texture to stop them sliding into infinity. Grant felt very comfortable in starkly simple environments. Who needed dazzling sunsets or intricate reconstructions of forests long since dead? Here there were just two parallel planes to provide the psychological shelter of a certain 'up' and 'down'. He hoped that Samantha wasn't too disappointed by the minimalism. Certainly, it was nice to be able to see her rendered to a megapolygon level.

Her hair was an intricate mass of soft, blonde fractal chaos that lay idly around her shoulders. Grant almost wished he'd programmed a breeze to watch the dynamic render to full effect. But then his VR-card was probably over-heating as it was.

The texture of her skin was incredible, and all interfaced for touch. There was a warmth to her breath in addition to that of her body, and an oft-forgotten clear membrane over her eyeballs that made her pupils sparkle and her bitmapped irises shine. Her lipstick and other make-up had probably been hand-painted in real-time to get the near-perfect but not digitally-antiseptic effect. Every nail carried a different, psychedelic video image. But then that was hardly difficult to do.

She was wearing that seamless, patent black jumpsuit that hugged every contour and felt like silk. Grant remembered how well the garment reflected a complex environment, and almost wished he *had* opted for a cloud-strewn sky. But then life was all about processing-capacity trade-offs, and he still preferred to lavish all available terahertz on their own bodies. Maybe one day they would both be able to afford consoles that would allow them to stroll and love full-resolution in any complexity of environment. Such a dream could come true, though only if they got good enough degrees. Which reminded him, they were supposed to be revising.

Aware of but ignoring his attention, Samantha was flipping through a personal video window. Gentle, almost playful motions of her right hand advanced the information flow. Math. Grant hated that. He was a programmer, and had little time for the abstraction of anything but object code. His mind started to wander again. Samantha stretched and settled into a more comfortable position. In megapolygon she was sensational. Though at times Grant did wonder what her real body was like.

With a sigh he thought into view his own video window and perused the timetable for the next semester after the exams. He hoped they could share as many classes as possible. However, he would have to study long and hard to remain in Samantha's league. Like most women, she was extremely conscientious and intelligent. One of his philosophy professors claimed that VR had freed the female spirit. Grant briefly glanced to his girlfriend, catching a distorted but realtime reflection of his face in her jumpsuit, and decided that whatever VR had done, a life without it must have been dire.

MindScrolling down the timetable, he came across a curious entry. It was a class that involved gathering in a common, physical location. The number of places was limited, and not even text links were allowed. He tapped Samantha on the shoulder to ask if she would like to jointly enrol. As he zoomed the window and pointed she just smiled. The class was called 'flesh socialization'. They both double-clicked.

Further Reading

Whilst the full references and other endnotes follow, below is a list of some of those books that I have most enjoyed over the past few years, and which I would most heartily recommend.

Being Digital, Nicholas Negroponte (London: Hodder & Stoughton, 1995).

Beyond Certainty: The Changing World of Organizations, Charles Handy (London: Hutchinson, 1995).

Chaos & Cyberculture, Timothy Leary (Berkeley, CA: Ronin, 1994).

*Children of Chaos > * [Surviving the End of the World as We Know It]*, Douglas Rushkoff (London: HarperCollins, 1997).

Computers as Theatre, Brenda Laurel (Reading, MA: Addison-Wesley, 1993).

Connexity: How to Live in a Connected World, Geoff Mulgan (London: Chatto & Windus, 1997).

Cyberia: Life in the Trenches of Hyperspace, Douglas Rushkoff (London: HarperCollins, 1994).

The Digital Economy: Promise and Peril in the Age of Networked Intelligence, Don Tapscott (New York: McGraw-Hill, 1995).

The Empty Raincoat: Making Sense of the Future, Charles Handy (London: Arrow, 1995).

The Fourth Discontinuity: The Co-evolution of Humans and Machines, Bruce Mazlish (New Haven: Yale University Press, 1993).

Glimpses of Heaven, Visions of Hell: Virtual Reality and its Implications, Barrie Sherman and Phil Judkins (London: Coronet, 1993).

Managing Generation X: How to Bring Out the Best in Young Talent, Bruce Tulgan (Capstone, Oxford: 1996).

Metaman: Humans, Machines and the Birth of a Global Super-organism, Gregory Stock (London: Bantam Press, 1993).

The Mode of Information, Mark Poster (Cambridge: Polity Press, 1990).

Net Gain: Expanding Markets Through Virtual Communities, John Hagel III & Arthur G. Armstrong (Boston, MA: Harvard Business School Press, 1997).

Resisting the Virtual Life: The Culture and Politics of Information, James Brook & Iain A. Boal (eds.) (San Francisco: City Lights, 1995).

The Road Ahead, Bill Gates with Nathan Myhrvold and Peter Rinearson (New York: Penguin, 1995).

The Search for Meaning, Charles Handy (London: Lemos & Crane, 1996).

Technotrends: How to Use Technology to Go Beyond Your Competition, Daniel Burrus, with Roger Gittines (New York: HarperCollins, 1993).

The Virtual Community: Finding Connection in a Computerized World, Howard Rheingold (London: Secker & Warburg, 1994).

The War of Desire and Technology at the Close of the Mechanical Age, Allucquère Rosanne Stone (Cambridge, MA: MIT Press, 1995).

War of the Worlds: Cyberspace and the High-tech Assault on Reality, Mark Slouka (London: Abacus, 1996).

References & Notes to All Chapters

All world-wide web references included herein were last accessed and verified as of 1st October 1998. For the latest and updated on-line links and other materials related to this book, see ***www.CREaction.demon.co.uk/value.html***

Chapter 1: Prelude

1. Thomas A. Stewart 'The Information Age in Charts', *Fortune International* (April 1994): 55.
2. A hard disk is the device mounted inside a computer that is most commonly used to store programs and user data files.
3. Cyberspace is the electronic data realm conceptualized to exist inside and across all computers and telecommunications networks.
4. One of the great 'problems' with the term the 'Industrial Revolution' is pinning down an accepted timespan. This clearly varied not just between nations, but between different regions thereof. Probably the most commonly accepted dates are between around 1760 to 1830.
5. For a far broader summary of this key managerial debate, *see* Sumantra Ghoshal and Christopher A. Bartlett 'Changing the Role of Top Management: Beyond Structure to Process', *Harvard Business Review* (January–February, 1995).
6. Just two of the early proponents of structure-led management were Henry Ford (who built the first mass production car plants), and Frederic Winslow Taylor (who championed 'scientific management'). *See* **chapter 5** pages 96-98.
7. Just two such process-led initiatives have been 'total quality management' (TQM) and 'business process reengineering' (BPR). For more details *see* **chapter 5**, pages 104–107.
8. Michael Porter's classic tomes include *Competitive Strategy* (New York: The Free Press, 1980) and *Competitive Advantage* (New York: The Free Press, 1985).
9. For a summary of recent work on virtual organizations *see*, for example, my earlier books *Cyber Business: Mindsets for a Wired Age* (Chichester: John Wiley & Sons, 1995): chapter 3, or *Challenging Reality: In Search of the Future Organization* (Chichester: John Wiley & Sons, 1997): chapter 16.
10. Deoxyribonucleic acid (DNA) is the very building block of life, with each DNA molecule encoded with the billions of bits of chemical information that determine each individual's physical make-up. The nucleus of every one of our body cells is thought to contain a single DNA molecule in the shape of a double helix which, if unwound, would be sixteen inches long.
11. Gregory Stock *Metaman: Humans, Machines & the Birth of a Global Super-Organism* (New York: Bantam Press, 1993).
12. Perhaps most notably, the concept of digital convergence has been championed by Nicholas Negroponte, Director of the Media Lab at the Massachusetts Institute of Technology.

Chapter 2: Networks & Middleware

13. One of the first individuals to write of the shift from 'personal' to 'interpersonal' computing was Timothy Leary in *Chaos & Cyberculture* (Berkeley, CA: Ronin, 1994).

14. The network of the Defence Advanced Research Project's Agency (DARPA—though now ARPA).

15. Amazon.com is a book store offering a catalogue of over 2.5 million volumes. The Innovations Group sells innovative consumer products in areas from personal care to DIY, gardening, textiles, housewares, sports and leisure. Value-Direct sells a range of 2500 electrical goods on-line, including televisions, washing machines and ovens. Now offering full on-line shopping, Tesco is one of the UK's largest supermarkets. *See* respectively **www.amazon.com**, **www.innovations. co.uk**, **www.value-direct.co.uk** and **www.tesco. co.uk/homeshop/**

16. The UPS web site is located at **www.ups.com/tracking/tracking.html**, whilst figures for its cost advantages are cited from *The Money Programme* (London: British Broadcasting Corporation, 12th October 1997).

17. It should be noted that there is a subtle but important difference between user authorization (where somebody is given the *rights* to access a system) and user authentication (where an authorized user is *identified* by a system as having such rights).

18. For example, as reported by Geoffrey Nairn in the *Financial Times Review of Information Technology* on the 4th of February 1998, a CMG survey of 250 UK companies found that more than 70% had an intranet. Of these, 94% were also planning to expand their intranet application.

19. The Netscape Communications Corporation, for example, define crossware as encompassing all of those on-demand applications that run across (most) networks and operating systems, and which are based entirely on open Internet standards like HTML, Java, and JavaScript.

20. Alex Summersby 'The Man at Apple's Core', *Macformat* (Issue 4: December 1996).

21. Marc Andreessen and the Netscape Product Team *The Networked Enterprise: Netscape Enterprise Vision and Product Roadmap*, Netscape Communications Company, 1997.

22. Don Tapscott *The Digital Economy: Promise and Peril in the Age of Networked Intelligence* (New York: McGraw-Hill, 1995): 56.

23. Ibid.

24. Oscar H. Gandy Jr. 'It's Discrimination, Stupid!' in James Brook & Iain A. Boal (eds) *Resisting the Virtual Life: The Culture and Politics of Information* (San Francisco: City lights, 1995): 39.

25. A 'cookie' is a small packet of data sent to a user's PC when they interact with certain web sites. Cookies permit web browsers to retain potentially useful, user-specific information. For example, cookies can store a user's ID and password, the details of previous searches, and may permit web page personalization. Highly targeted web tracking and web marketing thereby become possible. *See also* **www.cookiecentral.com**

26. For a more extensive investigation of actual and potential developments in the business application of virtual reality, *see* Christopher Barnatt *Cyber Business: Mindsets for a Wired Age* (Chichester: John Wiley & Sons, 1995).

27. A language known as VRML—the virtual reality modelling language—permits the creation and communication of three-dimensional, virtual reality worlds on the world-wide web, within which users may freely roam.

28. Several research initiatives are already developing the VR tools required to immerse remote participants in a single, virtual environment. For example, the four year European Union Advanced Communications COVEN project was launched in 1995 to develop COllaborative Virtual ENvironments that would 'support future cooperative teleworking systems and to demonstrate the added value of networked VR for both professional users and home users'. *See also* **http://chinon.thomson-csf.fr/projects/coven/**

29. National Aeronautics and Space Administration *Software for the First New Millennium Mission Closest Yet to 'HALL 9000'*, Press Release 1997 Number 7.

30. As reported by Paul Taylor in the *Financial Times Review of Information Technology* (6th March 1996): 1.

31. Alex Summersby, op.cit.: 29.

32. For further information on software agents *see* Christopher Barnatt 'Our New Working Class: The Business Implications of Software Agents', *Journal of General Management* (Volume 23 No.2: Winter 1997), or consult the UMBA Agents Web at **www.cs.umbc.edu/agents/**

33. Alongside Apple, the backers of General Magic are AT&T, Motorola, Philips, Matsushita and Sony.

34. Don Tapscott, op.cit: 112.

35. Stewart Brand *The Media Lab: Inventing the Future at MIT* (New York: Penguin, 1988): 9.

36. Charles B. Wang *Techno Vision* (New York: McGraw Hill, 1994).

37. Ibid.: 189.

38. I first introduced the concept of the global hardware platform in *Cyber Business: Mindsets for a Wired Age* (Chichester: John Wiley & Sons, 1995).

39. The transition from trade based upon the exchange of atoms, to trade instead based upon the exchange of computer data (or 'bits'), is explored in detail by Nicholas Negroponte within his book *Being Digital* (London: Hodder & Stoughton, 1995).

40. Don Tapscott, op.cit.: 220.

41. Geoffrey Robinson, 'Technology Foresight—The Future for IT', *Long Range Planning* (Volume 29, No.2. April 1996): 237–238.

42. Thomas H. Davenport and James E. Short 'The New Industrial Engineering: Information Technology and Business Process Redesign', *Sloan Management Review*, (Summer 1990): 12.

43. Ibid.

44. Bill Gates with Nathan Myhrvold and Peter Rinearson *The Road Ahead* (NY: Penguin Books, 1995): 5–6.

45. Ibid.: 158.

46. Ibid.

47. Most frequently, such a form of organization is termed an 'organic' or 'dynamic' network—a definition perhaps most notably attributed to this type of configuration by Raymond E. Miles and Charles C. Snow in their paper 'Organizations: New Concepts for New Forms' *California Management Review* XXVIII (Spring 1986): 62–73.

48. For a broader discussion of HCI, *see*, for example, Jenny Preece (ed) *A Guide to Usability: Human Factors in Computing* (Wokingham: Addison-Wesley, 1993), or Brenda Laurel *Computers as Theatre* (Reading, MA: Addison-Wesley, 1993).

49. For a broader analysis of the disconnect between business management and information technology, *see* Charles Wang, op.cit.

50. N. Caroline Daniels discusses the role of hybrid managers in chapter 8 of her book *Information Technology: The Management Challenge* (Wokingham: Addison-Wesley, 1994).

Chapter 3: Flexibility or Identity

51. For a far broader examination of the transition of our dominant 'member status' from serfs and slaves to employees to free agents, *see* Christopher Barnatt *Challenging Reality: In Search of the Future Organization* (Chichester: John Wiley & Sons, 1997): 63–100.

52. John Atkinson *The Flexibility Factor* (Sussex University: Institute for Manpower Studies, 1985).

53. Zero-hours contracts exist where staff remain on an employment register, ready to work as required, yet with no guarantee that any work will be available.

54. For more background on Charles Handy's work in this area, *see* his books *The Empty Raincoat* (London: Hutchinson, 1994) and *Beyond Certainty: The Changing World of Organizations* (London: Hutchinson, 1995).

55. The Institute of Management and Manpower Plc *Survey of Employment Strategies 1996: The 5th Annual Survey among leading UK organizations from the public, private and voluntary sectors*.

56. 'Realizing the Potential of Self-Employment', *Small Business Foresight Bulletin*, Issue 7, Durham University Business School in association with Nat West, 1997. Available at **www.dur.ac.uk/foresight/bulletin/bull_97/b7_97.html**

57. Manpower Inc and Manpower Europe *Argus: How Employment Looks, Worldwide: A monthly digest of worldwide employment news*. November 1997. Further information on Manpower Reports can be found at **www.manpower.co.uk**

58. Ibid.

59. Dom Pancucci 'Remote Control', *Management Today* (April 1995): 78.

60. Charles Handy discusses in depth our emerging portfolio workstyle in his books as referenced in note 54.

61. For example, more than 5 million people in Europe (3.5% of the workforce) now have two or more jobs. *Source*: Manpower Inc and Manpower Europe, op.cit.

62. 'Realizing the Potential of Self-Employment', op.cit.

63. It ought, perhaps, to be noted here that religions have always traded in community affiliation. However, few would class religions as 'markets', let alone as a 'growth industry'.

64. The battle is hotting up between those software companies seeking to sell their virtual community software as a new market standard. Current players include Durand Communications with *CommunityWare* (*see* **www.communityware.com**), Screen Porch with *Caucus 4.0* (*see* **www.screenporch.com**), and Well Engaged with *Discussions 1.5* (*see* **www.wellengaged.com**).

65. John Hagel III & Arthur G. Armstrong *Net Gain: Expanding Markets through Virtual Communities* (Boston, MA: Harvard Business School Press, 1997).

66. Ibid.: 29.

67. Mark McDonough *Frequently Asked Questions: Virtual Communities*. Internal paper prepared for virtual community hosts at International Thomson's Virtual Community Labs, 1997.

68. John Hagel III & Arthur G. Armstrong, op.cit. 66.

69. Ibid.: 75.

70. Ibid.: 216.

71. For a broader discussion of virtual community creation versus inhabitation strategies, *see* Christopher Barnatt 'Virtual Communities and Financial Services: On-line Business Potentials and Strategic Choice', *International Journal of Bank Marketing* (Volume 16, Number 4: 1998).

72. Geoff Mulgan *Connexity: How to Live in a Connected World* (London: Chatto & Windus, 1997).

Chapter 4: Playing at God

73. Cited from James Paradis *T.H. Huxley: Man's Place in Nature* (Lincoln: University of Nebraska Press, 1978): 150.

74. Bruce Mazlish *The Fourth Discontinuity: The Co-evolution of Humans and Machines* (New Haven: Yale University Press, 1993).

75. Ibid.: 6.

76. Many retailers now have their stock control systems networked with electronic data interchange (EDI) supplier links. Across such systems, orders for fresh goods may be automatically placed based on aggregated retail data from actual sales.

77. One may cite here manufactured devices ranging from spectacles and contact lenses, to deaf aids, cochlear implants, pacemakers, artificial limbs, and perhaps even voice-controlled computing devices.

78. Whilst a standard computer mouse is moved over a flat surface to move a pointer on the computer screen, a trackball is an 'upside-down mouse' that remains static whilst a user rolls its top-mounted rollerball. Touchpads, tablets and touch-sensitive screens allow computer control by pointing or drawing with a finger, stylus or puck.

79. It is, for example, now possible to dictate into a computer and to control *Windows* with voice commands. Similarly, it is commonplace to use a scanner to capture an image of a document which OCR (optical character recognition) software may subsequently (if not totally accurately) translate into a manipulable data format.

80. *See* note 3.

81. A dataglove is a computer peripheral worn on the hand to permit the movement of the user's hand and fingers to be mirrored in a virtual reality graphics world.

82. Mark Billinghurst, Suzanne Weghorst & Tom Furness III (1997) *Wearable Computers for Three Dimensional CSC* (Human Interface Technology Lab, University of Washington, 1997): 1. Available at **www.hitl.washington.edu/projects/wearables/papers.html**

83. For example, Ford has for several years prototyped new vehicles in VR, whilst supermarkets Sainsburys and Tesco have used VR to plan and model new store layouts.

84. With VR imaging systems fed from real patient scan data, surgeons may finally be freed from the limitations of planning operations from 2-D X-ray pictures. VR medical systems can also allow doctors to practice complex procedures before cutting flesh.

85. A 'virtual office' and a 'virtual catwalk' were both developed under the auspices of the *Virtuosi Project*, a three-year initiative led by British Telecom and funded by the SERC and the DTI CSCW programme (UK). Other partners included leading European VR systems supplier Division, GPT and the GEC-Marconi Research Centre, BICC Ltd, Nottinghamshire County Council, and researchers at the Universities of Nottingham, Lancaster and Manchester. For further information *see* Benford, S., Bowers, J., Gray, S. Roden, T., Ryan, G. and Stanger, V. 'The Virtuosi Project' in *Proceedings of VR 94*, held in London as part of Virtual Reality Expo (February 1994). *See also* note 28.

86.   E. Brodie 'Virtual Reality Takes Fund Managers into Cyberspace' *The Independent on Sunday* (1st August 1993).

87.   *See* Dom Pancucci 'The Real Thing' *Which Computer* (August 1993): 38–43.

88.   John Casti *Would-Be Worlds: How Simulation is Changing the Frontiers of Science* (New York: John Wiley & Sons, 1997): 35.

89.   Mark Billinghurst, Suzanne Weghorst & Tom Furness III, op.cit: 1.

90.   For example, in the March of 1998, the bright-yellow *VRD* PC virtual reality headset was launched from Virtuality for $799. Plug-and-play compatible with a standard 3D PC graphics card, the VRD weighs about 500 grams, and tracks head movement at 256Hz to permit fluid VR immersion with minimal jitter and lag. Based on the same 225×800 optics system from Retinal Displays Inc., Philips *Scuba* headset also recently became available for game consoles for $299.

91.   *Cited from* the television documentary *Future Fantastic* on virtual reality produced by Jasper James and directed by Samantha Starbuck (London: BBC TV in co-production with The Learning Channel and Pro Sieben, 1996). Series producer David McNab.

92.   Information on the virtual retinal display (VRD) project funded by Microvision Inc. at the Human Interface Technology Lab from **www.hitl.washington.edu/projects/vrd/**

93.   As explained by Peter Cochrane in a lecture delivered at the *BIT 94: Virtual Reality Applications & Implications* conference, University of Leeds, 24th March 1994.

94.   Timothy Leary *Chaos and Cyberculture* (Berkeley, CA: Ronin, 1994): 20.

95.   Ibid.

96.   One startling implication of being able to link mindspace to cyberspace, and hence one mindspace to another via cyberspace, is that human beings may evolve the ability to think in and as group entities. However, almost by definition, such a concept has to lie beyond the comprehension of beings like ourselves who have always lived in single minds.

97.   The Dobelle Institute was established in 1993 by Dr. William H. Dobelle. In addition to developing artificial vision systems, it currently manufactures several types of clinical device based on a man–machine interface. These include pacemakers; neuro-stimulation devices for implantation into the brain, spinal cord, or peripheral nervous system to suppress chronic pain; urogenital implants that control voiding, incontinence, or male sexual dysfunction; deep brain implants for movement disorders; and cerebellar implants for cerebral palsy and epilepsy. *See* **http://cpmcnet.cpmc.columbia.edu/news/audubon/archives/audu_v1n1_0007.html**

98.   For a slightly broader discussion of direct human–computer interface possibilities and developments, *see* Christopher Barnatt, *Cyber Business: Mindsets for a Wired Age* (Chichester: John Wiley & Sons, 1995): 135–139.

99.   As noted by Robert Winston in *The Future of Genetic Manipulation* (London: Phoenix, 1997) ancient Vedic, Greek and Talmudic literature all contain references to attempted sex selection, as do many Biblical tales.

100. The Human Genome Project is run under the auspices of the Human Genome Organization (HUGO), and at the time of writing involved the collaboration of formal genome programmes in the United States, Japan, the former Soviet Union, Denmark, France, Germany, Italy and the United Kingdom.

101. *See* note 10.

102. Robert Winston *The Future of Genetic Manipulation* (London: Phoenix, 1997): 26.

103. Ibid.: 32.

104. The cloning technique pioneered at the Roslin Institute involved sucking the DNA out of an infertile sheep's egg. This genetically barren egg was then 'reprogrammed' by implanting it with alternative sheep DNA derived from a foetus, and which had previously been cultured (potentially limitlessly) within the laboratory. An electrical pulse was used to release the genetic material from the implanted cells so that they would 'take over' their new host egg. Finally, the reengineered egg was implanted into a surrogate mother within whose womb it could develop to maturity.

105. Robert Winston, op.cit.: 35.

106. It is perhaps interesting to note that, in the United States, there are already hundred-million-dollar fertility companies running 'reproductive science centres' that trade in conception and reengineered embryos—and which even offer money-back guarantees. *Source*: *Channel 4 News* (London: Channel 4 Television plc, 5th December 1997).

107. William Day *Genesis on Planet Earth: The Search for Life's Beginning* (East Lansing, Mich.: House of Talos, 1979): 390–392.

108. Bruce Mazlish, op.cit: 174.

109. Ibid.: 175—taking a term also used by Freud.

110. Ibid.: 161.

111. A nanometre is one thousand-millionth of a metre.

112. Eric Drexler's publications on nanotechnology include *Engines of Creation: The Coming Era of Nanotechnology* (Anchor/Doubleday, 1986) and (with Chris Peterson and Gayle Pergamit, 1991) *Unbounding the Future: The Nanotechnology Revolution*. Both of these books, in addition to a wide variety of other nanotechnology materials, are available from **www.foresight.org/NanoRev**. For the more general reader, Ed Regis' book *Nano!* (London: Bantam, 1995) also comes highly recommended.

113. Lithography is a photographic technique used in the production of silicon chips, and whose miniaturization is approaching its optical limits.

114. Reports and examples cited from Vanessa Houlder 'Matters of Scale' *Financial Times* (Thursday 4th September 1997).

115. Colossus was created by Alan Turing and his colleagues at Bletchley Park in England, and relied upon high-speed lengths of punched paper tape, photoelectric cells, and 15,000 vacuum tubes, in its descrambling of coded messages. By the end of WWII, ten such top secret computers—or Colossi—had been brought into military service.

116. Doctor Who *The Curse of Fenric* by Ian Briggs (London: BBC Television, 1989): Episode 1.

117. *Cited in* Barry Sherman and Phil Judkins *Glimpses of Heaven, Visions of Hell: Virtual Reality and its Implications* (London: Hodder & Stoughton, 1993): 188.

118. *Cited from* Sheila Hayman *The Electronic Frontier*, transcript of BBC *Horizon* television documentary (London: BBS, 1993): 6.

119. The concept of 'digital nomads'—of people free to roam in geography and time due to the emergence of a lightweight, low cost 'Complete Nomadic Toolset' comprising all required computing and communications technologies—is discussed by Tsugio Makimoto and David Manners in their book *Digital Nomad* (Chichester: John Wiley & Sons, 1997).

120. Both this definition of transhumanity, and the previous one by Anders Sandberg, are cited from Anders Sandberg's *Definitions of Transhumanism* world-wide web page, located at **www.aleph.se/Trans/Intro/Definitions.html**

121. Some transhumanists now pay to have their whole body, or sometimes just their head and spinal chord, frozen for possible 'reanimation' once medical science has

sufficiently advanced. For transhumanists, cryogenic freezing is also potentially seen as a means of enabling human beings to survive long journeys into space.

122. The concept of 'uploading' at present remains pure fiction. However, a wide range of literature on this area can be found at **www.aleph.se/Trans/Global/Uploading**

123. The idea of compiling a list of the principles of transhumanity was put forward by Alex Bokov. Other major contributors to the discussions that resulted in the list as reproduced in the text include Anders Sandberg, Rich Artym, Nancie Clark, Romana Machado, Sasha Chislenko, Mark A. Plus and Christopher T. Brown. The full archive of the discussions that lead to the drafting of the 'consensus list' is available at **www.lucifer. com/~sasha/refs/Principles_Archive.html**

124. Bruce Mazlish, op.cit.: 201.

125. We may perhaps reflect here upon the encoding of human beings into their software agents, which may potentially lead to the creation of 'software ghosts' of human personalities roaming immortal in cyberspace. For more discussion *see* Christopher Barnatt *Cyber Business: Mindsets for a Wired Age* (Chichester: John Wiley & Sons, 1995): chapter 4.

126. Timothy Leary, op.cit. 202.

Chapter 5: Maximising Corporate Success

127. For a more extended discussion of the development and principles of systematic, scientific and administrative management, *see* (for example) Thomas S. Bateman and Carl P. Zeithaml *Management Function & Strategy*, 2nd Edition (Homewood, IL: Irwin, 1993): 32–40.

128. For a broader exploration of this trend, *see* Christopher Barnatt *Challenging Reality: In Search of the Future Organization* (Chichester: John Wiley & Sons, 1997): chapter 6.

129. In return for working in Taylor's one best way at Bethlehem, 'high priced men' were rewarded with 60% higher wages. For more information upon Taylor's experiments at the Bethlehem steel works and elsewhere, *see* F.W. Taylor *Scientific Management* (New York: Harper & Row, 1947).

130. Henri Fayol's 14 principles of management were a specialized division of work; the delegation of authority; clear discipline; a unity of command for workers; a unity of direction for the organization; the subordination of individual interest to the general interest; systematic worker remuneration; centralization; the establishment of a scalar chain of command; order to support the organization's direction; a sense of equity to enhance employee commitment; stability and tenure of personnel; encouragement of initiative; and an 'esprit de corps' to promote a unity of interest between employees and management.

131. For a fuller account of the Hawthorne Studies, *see* Elton Mayo 'Hawthorne and the Western Electric Company' in D.S. Pugh (ed) *Organization Theory: Selected Readings* (London: Penguin, 1990): 345–357.

132. The five progressive levels of Maslow's hierarchy are physiological survival needs (eg for air, food and water); safety needs (eg for shelter); belongingness needs (for social relationships); ego-status needs (for improved self-esteem); and finally the somewhat dubious level of self-actualization (or total personal fulfilment).

133. The full account of E.A. Trist & K.W. Bamforth's study of 'Some Social and Psychological Consequences of the Longwall Method of Coal Getting' can be found in *Human Relations* (Volume 4, Number 1, 1951): 6–24 & 37–38.

134. For a full account of this study, *see* Tom Burns and G.M. Stalker *The Management of Innovation* (London: Tavistock, 1961).

135. Perhaps the two most important studies here were those undertaken by Peter R. Lawrence & Jay W. Lorsh in 1967 (*see*, 'High Performing Organizations in Three Environments' reproduced in D.S. Pugh (ed) *Organization Theory: Selected Readings* (London: Penguin, 1990), and by Joan Woodward, who concluded in *Industrial Organization: Theory and Practice* (London: Oxford University Press, 1965) that particular forms of organization were most appropriate to different degrees of technological process complexity.

136. Thomas J. Peters & Robert H. Waterman Jr. *In Search of Excellence: Lessons from America's Best-Run Companies* (New York: Harper & Row, 1982).

137. Ibid.: 3.

138. Ibid.: 8. A broader discussion of the strategy–structure debate can also be found in **chapter 1** of this book on pages 11–12.

139. Thomas J. Peters & Robert Waterman, op.cit.: 9–11.

140. Ibid.: 13.

141. G. Hofstede, B. Neuijen, D. Ohayv and G. Sanders 'Measuring Organizational Cultures: A Qualitative Study Across Twenty Cases' *Administrative Science Quarterly* (Volume 35, 1986): 286–316.

142. Andrew Brown *Organizational Culture* (London: Pitman, 1995): 10.

143. Gerald Egan 'Cultivate Your Culture' *Management Today* (April 1994).

144. This description of a corporate culture is attributable to Gareth Morgan in *Creative Organization Theory: A Resourcebook* (Newbury Park, CA: Sage Publications, 1989): 157.

145. Just-in-time manufacturing demands that stock-levels are kept to a minimum by ensuring that components are delivered by suppliers to the production line only as and when required. JIT hence eliminates waste and reduces process costs, and is a feature of many successful Japanese manufacturing processes.

146. For a summary of Deming's 'fourteen points' of TQM, *see* Andrezej A. Huczynski *Management Gurus: What Makes Them and How to Become One* (London: International Thomson Business Press, 1996): 183.

147. Michael Hammer & James Champy *Reengineering the Corporation: A Manifesto for Business Revolution* (London: Nicholas Brealey, 1993): 32.

148. Christopher Barnatt *Management Strategy and Information Technology: Text and Readings* (London: International Thomson Business Press, 1996): 73.

149. Hammer and Champy, op.cit.: 83.

150. For a full account of the range of IT levers than may be utilised in a recursive BPR relationship, *see* Thomas H. Davenport and James E. Short 'The New Industrial Engineering: Information Technology and Business Process Redesign', *Sloan Management Review* (Volume 31, Number 4, Summer 1990).

151. Cited from the Hyundai web site at **www.hyundai.net**

152. Cited from the 'Value Solutions' *Value Based Management* pages of the Deloitte Touche Consulting Group web site at **www.dtcg.com/value_solutions/value_based_management01.htm**

153. Ibid.

154. Cited from the Center for Economic and Social Justice *Value Based Management: A System for Transforming the Corporate Culture* at **www.cesj.org/vbm3pgs.htm**

155. Ibid.

156. Ibid.

157. Figures cited from a report by The Millennium Group in February 1998 on 'Anticipating the Organizational Impact of Technological Change'—the fourth report from Millennium's 1997 *Exploiting New Opportunities to Create Value* research and development programme.

158. This distinction between 'knowledge' and 'information' is taken from the prequel to this book, *Challenging Reality: In Search of the Future Organization* (Chichester: John Wiley & Sons, 1997), and is explored in more detail within chapter 10 of the same.

159. As note 157.

160. Peter M. Senge *The Fifth Discipline: The Art and Practice of the Learning Organization* (London: Random House, 1995).

161. David Birchall and Lawrence Lyons *Creating Tomorrow's Organization: Unlocking the Benefits of Future Work* (London: Pitman, 1995).

162. As note 157: 32.

163. Daniel Burrus with Roger Gittines *Technotrends: How to Use Technology to Go Beyond Your Competition* (New York: Harper Business, 1993): 322–323.

Chapter 6: Voices from Cyberspace

164. From *You Were So Far From Me* by Vivien Steels, reproduced with the permission of the author.

165. According to statistics from the Department of Trade and Industry's *Information Society Initiative*, over twenty per cent of non-owners see PCs and the Internet as 'useful'.

166. Myron W Krueger *Artificial Reality II* (Reading, MA: Addison-Wesley, 1991): xi.

167. William Gibson *Neuromancer* (New York: Ace Books, 1984): 67.

168. William Gibson *Mona Lisa Overdrive* (London: Victor Gollancz, 1988): 22.

169. Allucquère Rosanne Stone 'Will the Real Body Please Stand up?: Boundary Stories about Virtual Cultures' in Michael Benedikt (ed) *Cyberspace: First Steps* (Cambridge, MA: MIT Press, 1991): 99.

170. Allucquère Rosanne Stone *The War of Desire and Technology at the Close of the Mechanical Age* (Cambridge, MA: MIT Press, 1995): 34.

171. Allucquère Rosanne Stone 'Will the Real Body Please Stand up?: Boundary Stories about Virtual Cultures', op.cit.: 103.

172. Cited from Douglas Rushkoff *Cyberia: Life in the Trenches of Hyperspace* (London: Harper Collins, 1994): 71–72.

173. Mark Slouka *War of the Worlds: Cyberspace and the high-tech assault on reality* (London: Abacus, 1996): 29.

174. Ibid.: 34.

175. Ibid.: 68.

176. Rebecca Solnit 'The Gardens of Virtual Life', in James Brook & Ian A. Boal (eds) *Resisting the Virtual Life: The Culture and Politics of Information* (San Francisco: City Lights, 1995): 229.

177. Allucquère Rosanne Stone 'Will the Real Body Please Stand up?: Boundary Stories about Virtual Cultures', op.cit.: 85.

178. It is perhaps interesting to note here that, by percentage of population, university students are one of most wired groups of people on the planet.

179. Barry Sherman & Phil Judkins *Glimpses of Heaven, Visions of Hell: Virtual Reality and its Implications* (London: Hodder & Stoughton, 1993): 201.

180. Howard Rheingold *The Virtual Community: Finding Connection in a Computerized World* (London: Secker & Warburg, 1994): 151.

181. Allucquère Rosanne Stone *The War of Desire and Technology at the Close of the Mechanical Age*, op.cit: 27.

182. Ibid.: 87.

183. Mark Slouka, op.cit.: 131.

184. Stewart Brand *The Media Lab: Inventing the Future at MIT* (New York: Penguin, 1988): 249.

185. Ibid.: 255.

186. Myron Krueger, op.cit.: xiv.

187. John Hagel III & Arthur G. Armstrong *Net Gain: Expanding Markets through Virtual Communities* (Boston, MA: Harvard Business School Press, 1997): ix.

188. Timothy Leary *Chaos & Cyberculture* (Berkeley, CA: Ronin. 1994): 8.

189. Ibid.: 84.

190. Douglas Rushkoff *Cyberia: Life in the Trenches of Hyperspace* (London: Harper Collins, 1994): 52.

191. Nicholas Negroponte *Being Digital* (London: Hodder & Stoughton, 1995): 228.

192. Mark Poster *The Mode of Information: Post Structuralism and Social Context* (Cambridge: Polity Press, 1992): 72.

193. Barry Sherman and Phil Judkins; op.cit.: 232.

194. George Lakoff, interviewed by Ian A. Boal 'Body, Brain and Communication' in James Brook & Ian A. Boal (eds) *Resisting the Virtual life: The Culture and Politics of Information* (San Francisco: City Lights, 1995): 125.

195. Bill Gates with Nathan Myhrvold & Peter Rinearson *The Road Ahead* (NY: Penguin Books, 1995): 158.

196. Don Tapscott *The Digital Economy: Promise and Peril in the Age of Networked Intelligence* (New York: McGraw Hill, 1995): 92.

197. Bill Gates *et al*, op.cit.: 136.

198. Ibid.: 209.

199. Douglas Rushkoff, op.cit.: 276.

200. Timothy Leary, op.cit.: 75.

201. Mark Slouka, op.cit.: 119.

202. Bill Gates *et al*, op.cit.: 183.

203. Brenda Laurel *Computers as Theatre* (Reading, MA: Addison-Wesley, 1993): 214.

204. From an interview 'Messages from Babylon' with Mira Furlan by Joe Nazzaro in *The Official Babylon 5 Magazine* (Volume 1, Number 7, April 1998): 14.

205. Nicholas Negroponte, op.cit.: 178.

206. Stewart Brand, op.cit.: 33.

207. Douglas Rushkoff, op.cit.: 57.

208. Christopher Barnatt *Cyber Business: Mindsets for a Wired Age* (Chichester: John Wiley & Sons, 1995).

209. Don Tapscott, op.cit.: xiv.

210. Ibid.: xiv-v.

211. Allucquère Rosanne Stone 'Will the Real Body Please Stand Up?: Boundary Stories about Virtual Cultures', op.cit.: 101.

212. Stewart Brand, op.cit.: 242.

213. Kevin Robins & Les Levidow 'Soldier, Cyborg, Citizen' in James Brook & Iain A. Boal (eds) *Resisting the Virtual Life: The Culture and Politics of Information* (San Francisco: City Lights, 1995): 105–106.

214. Mark Slouka, op.cit.: 95.

215. Kenichi Ohmae 'Putting Global Logic First' *Harvard Business Review* (January–February 1995): 119–124.

Chapter 7: The neXt Generation?

216. Douglas Rushkoff *Children of Chaos* > * *[Surviving the End of the World as We Know It]* (London: HarperCollins, 1997): 13.

217. Ibid.: 2.

218. Ibid.

219. Geoff Mulgan *Connexity: How to Live in a Connected World* (London: Chatto & Windus, 1997): 44.

220. Bruce Tulgan *Managing Generation X: How to Bring Out the Best in Young Talent* (Oxford: Capstone Publishing, 1996): 18.

221. Ibid.: 32.

222. Ibid.: 3.

223. Ibid.: 4.

224. Ibid.: 14.

225. Tsugio Makimoto & David Manners *Digital Nomad* (Chichester: John Wiley & Sons, 1997): 8.

226. Ibid.: 9–10.

227. Ibid.: 3.

228. According to John Naisbitt in *Global Paradox* (New York: Avon Books, 1995), travel is already the world's largest industry in terms of gross output, and in 1994 was valued at around $3.4bn per annum.

229. For a detailed exploration of likely developments in space travel that look destined to broadly extend humanity's geographic span, *see* Christopher Barnatt *Challenging Reality: In Search of the Future Organization* (Chichester: John Wiley & Sons, 1997): chapter 13.

230. CND is the Campaign for Nuclear Disarmament. MTV is Music Television, a twenty-four-hour, global pop video entertainment channel.

231. Bruce Tulgan, op.cit.: 25–26.

232. For more than ten years compact disks (CDs) have permitted programmable, random access to music tracks and—more recently—computer data. Digital versatile disks (also known in some quarters as digital video disks) similarly now permit random access to digital video materials without having to wind back and forth spools of tape in order to locate the desired section of a movie or other programme.

233. For a broader exploration of multimedia to metamedia progression *see* Christopher Barnatt *Challenging Reality: In Search of the Future Organization* (Chichester: John Wiley & Sons, 1997): chapter 10. *See also* **chapter 8** of this book, pages 164–165.

234. Douglas Rushkoff, op.cit.: 44.

235. Ibid.: 49.

236. Ibid.: 39.

237. Geoff Mulgan, op.cit.: 44.

238. Bruce Tulgan, op.cit.: 25.

239. Ibid.: 37–38.

240. Douglas Rushkoff, op.cit.: 36.

241. Reported from the documentation prepared for the 'New Markets, New Approaches, New Segments' workshop run by the Millennium Group within their 1998 Research Programme *Creating Value with Technology*.

242. Ibid.

243. Ibid.

244. *Prepared for the Future? The British and Technology*. Motorola Report 1996, based on research carried out by MORI for Motorola July to September 1996.

245. For a fuller exploration of the Future Mindset, *see* Christopher Barnatt *Challenging Reality: In Search of the Future Organization* (Chichester: John Wiley & Sons, 1997): chapter 17.

246. Douglas Rushkoff, op.cit.: 269.

Chapter 8: Millennial Realities

247. Christopher Barnatt *Cyber Business: Mindsets for a Wired Age* (Chichester: John Wiley & Sons, 1995): 208.

248. Alvin Toffler *Future Shock* (London: The Bodley Head, 1970).

249. Alvin Toffler *The Third Wave* (London: Collins, 1980).

250. From 1972 the Japanese developed an HDTV system called Hi-Vision, whilst Europe opted for its own HD-MAC.

251. For a far more detailed coverage of the 'silliness' of HDTV development, *see* Nicholas Negroponte *Being Digital* (London: Hodder & Stoughton, 1996): 37–48.

252. Ibid.: 41.

253. The Five Facets model was first presented as the backbone to my book *Challenging Reality* (Chichester: John Wiley & Sons, 1997). The Holistic Lens model was developed in consultation with Ian McDonald Wood of the Millennium Group and Pat Lalonde-Dade of Synergy Consulting.

254. Michael J. Piore & Charles F. Sabel *The Second Industrial Divide: Possibilities for Prosperity* (New York: Basic Books, 1984).

255. Charles Handy *Beyond Certainty: The Changing World of Organizations* (London: Hutchinson, 1995): 31.

256. For a detailed exploration of this point of view, *see*, for example, Mark Poster *The Mode of Information: Post Structuralism and Social Context* (Cambridge: Polity Press, 1992).

257. Malcolm Waters *Globalization* (London: Routledge, 1995): 9.

258. Christopher Barnatt *Challenging Reality: In Search of the Future Organization* (Chichester: John Wiley & Sons, 1997): 151.

259. The debate here is inevitably complex and can hardly be done justice to in a sentence. Interested readers—and/or those disputing such claims—are therefore encouraged to consult my book *Challenging Reality*, op.cit.: chapters 11–13.

260. Raymond E. Miles & Charles C. Snow 'Organizations: New Concepts for New Forms' *California Management Review* XXVIII (Spring 1986): 62–73.

261. Don Tapscott *The Digital Economy: Promise and Peril in the Age of Networked Intelligence* (New York: McGraw Hill, 1995).

262. See note 9.

263. Critical to the success of any Holistic Lens brainstorming exercise is for all lists of outputs to be developed by workshop participants. If instead lists of WorldView and then PeopleView changes are presented as given, then the causalities between the different stages of the lens are unlikely to be brought to the fore. This said, some prompting by the facilitator from initial lists as presented in **chapter 8** can prove helpful with some groups.

264. The Selye-Toffler University is located at: **http://selye-toffler.org/index.html**

265. Ibid.

266. Geoff Mulgan *Connexity: How to Live in a Connected World* (London: Chatto & Windus, 1997): 109.

Chapter 9: Hopes & Fears for Century 21

267. Christopher Barnatt *Challenging Reality* (Chichester: John Wiley & Sons, 1997): 262.

268. It is perhaps important to note here that acts of self-sacrifice—at either an individual or organizational level—may significantly touch and influence others across great distances for long periods of time. Continued survival therefore need not *in all cases* be a prerequisite for human touch or even creation.

269. Don Tapscott *The Digital Economy: Promise and Peril in the Age of Networked Intelligence* (New York: McGraw Hill, 1995): 26.

270. PepsiCo Inc.'s unprecedented promotional alliance with Lucasfilm Ltd commenced in February 1997. It involves relationship tie-ins for exclusive, global on-package and in-package promotions within PepsiCo's brand categories for characters from the digitally remastered first *Star Wars* trilogy, as well as the first *Star Wars* prequel. For more information on this $2bn deal, *see* 'PepsiCo and Lucasfilm Unite for Largest Promotional Alliance in History', *The Official Star Wars Magazine* (August/September 1996): 4.

271. Cited from Stephen Edward Poe *A Vision of the Future: Star Trek Voyager* (New York: Pocket Books, 1998): 52.

272. Ibid.: 352.

273. Lester R. Brown, Michael Renner & Christopher Flavin *Vital Signs 1998: The Environmental Trends that are Shaping Our Future* (New York: Norton/Worldwatch Institute, 1998). Further information and summary materials from this report can also be found at **www.worldwatch.org/alerts/pr98507.html**

274. For an exploration of this more radical yet feasible option, *see* Christopher Barnatt, op.cit.: chapter 13.

275. Gregory Stock *Metaman: Humans, Machines, and the Birth of a Global Super-organism* (London: Bantam Press, 1993): 229.

276. Christopher Barnatt, op.cit.: 258.

277. In a similar fashion, information has little or no value in today's 'information age' due to its ubiquitous, mass availability. As the low value placed on air, not to mention traditional demand and supply economics, clearly demonstrate, no resource in a constant, abundant supply is ever likely to be attributed with great value. Any economy based on information or ideas is therefore extremely unlikely to value information or ideas themselves.

Index

ISBN 0-275-96714-X

90000>

HARDCOVER BAR CODE